# Introduction to R Programming Language

# Introduction to R Programming Language

**Mohsen Nady**

ARCLER PRESS

www.arclerpress.com

**Introduction to R Programming Language**

*Mohsen Nady*

**Arcler Press**
224 Shoreacres Road
Burlington, ON L7L 2H2
Canada
www.arclerpress.com
Email: orders@arclereducation.com



ISBN: 978-1-77469-039-0 (Hardcover)

# ABOUT THE AUTHOR

**Mohsen Nady** is a pharmacist with a M.D. in Microbiology and a diploma in Industrial Pharmacy. In addition, Mohsen has more than 4 years experience using R programming language. Mohsen has applied his skills in R programming to different projects related to Genomics, Microbiology, Biostatistics, Six Sigma, Data Analytics, Data Visualization, Building Apps, Geography, Market Analysis, Business Analysis,.....etc. Mohsen also published his thesis in high impact journal that attracted many citations, where all the statistical analysis were performed by him in addition to the methodological part. Furthermore, Mohsen has earned additional certificates, from top universities (Harvard, Johns Hopkins, Denmark,...etc) in R programming, Python, Excel, and Minitab that highlight his outstanding programming skills.

# TABLE OF CONTENTS

# LIST OF ABBREVIATIONS

| | |
|---|---|
| CRAN | Comprehensive R Archive Network |
| EST | Eastern Standard Time Zone |
| EWR | Newark International Airport |
| FHR | Fetal Heart Rate |
| GUI | Graphical User Interface |
| HTN | Hypertension |
| IDE | Integrated Development Environment |
| LGA | LaGuardia Airport |
| PUMS | Public Use Microdata Samples |
| UC | Uterine Contraction |

# PREFACE

This book covers some introductory steps in using R programming language as a data science tool. The data science field has evolved so much recently with incredible quantities of generated data. To extract value from those data, one needs to be trained in the proper data science skills like statistical analysis, data cleaning, data visualization, and machine learning. R is now considered the centerpiece language for doing all these data science skills because it has many useful packages that not only can perform all the previous skills, but also, has additional packages that was developed by different scientists in diverse fields. These fields include, but are not limited to, business, marketing, microbiology, social science, geography, genomics, environmental science, etc. Furthermore, R is free software and can run on all major platforms: Windows, Mac Os, and UNIX/Linux.

The first two chapters involve installing and using R and RStudio. RStudio is an IDE (integrated development environment) that makes R easier to use and is more similar to SPSS or Stata.

Chapters 3–8 covers the different R objects and how to manipulate them including the very popular one, dataframes. Chapter 9 is about importing different files into your R working session like text or excel files. Chapters 10 and 11 are dealing with different tidyverse packages that can do interesting summaries of different dataframes including different types of data visualizations.

In the last chapter, it introduces how functions are created in R along with some control structures and useful functions.

In all these chapters, many examples along with different codes and outputs are given to help your understanding of this powerful programming language. I hope this book will be great addition to your future data analysis projects.

CHAPTER 1

# INSTALLING R AND RSTUDIO

## CONTENTS

## 1.1 INSTALLING R

R is a software environment that comes with a graphical user interface (GUI). R GUI looks more similar to the old DOS console.

RStudio is an integrated development environment (IDE) that makes R easier to use and is more similar to SPSS or Stata. It includes a code editor, debugging, and visualization tools.

RStudio is not R, nor does it include R when you download and install it. Therefore, to use RStudio, we first need to install R. Installing R alone is similar to buying a car, while installing R and RStudio is like buying a car with all its accessories so both are important.

### 1.1.1 Steps

1. You can download R from the comprehensive R archive network (CRAN). Search for CRAN on your browser.

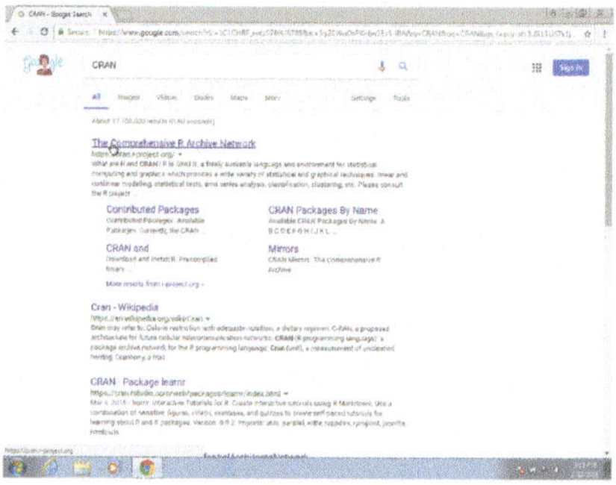

2. On the CRAN page, select the version for your operating system.

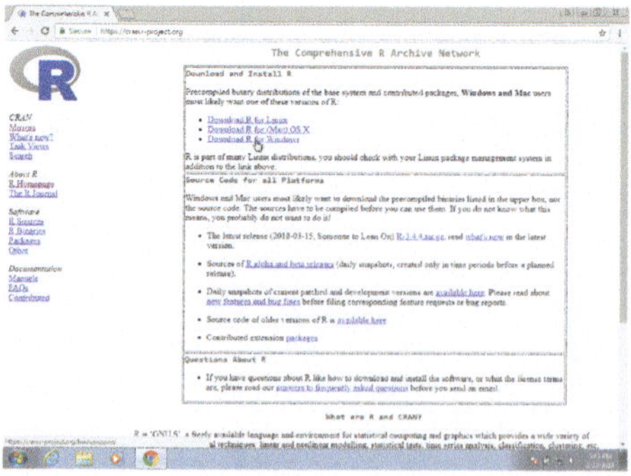

3. During installation, say yes to all defaults. This installs the basic packages you need to get started.
4. Congratulations! You have installed R.

## 1.2 INSTALLING RSTUDIO

### 1.2.1 Steps

1. You can start by searching for RStudio on your browser.

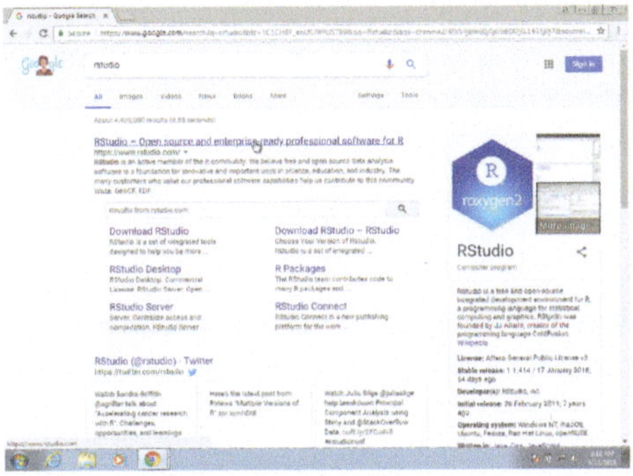


2. You should find the RStudio website as shown above. Once there, click on Download RStudio.

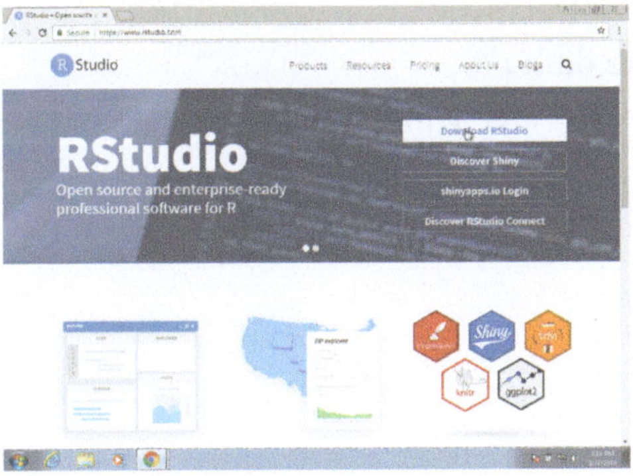


3. This will give you several options. Use the free Desktop version.

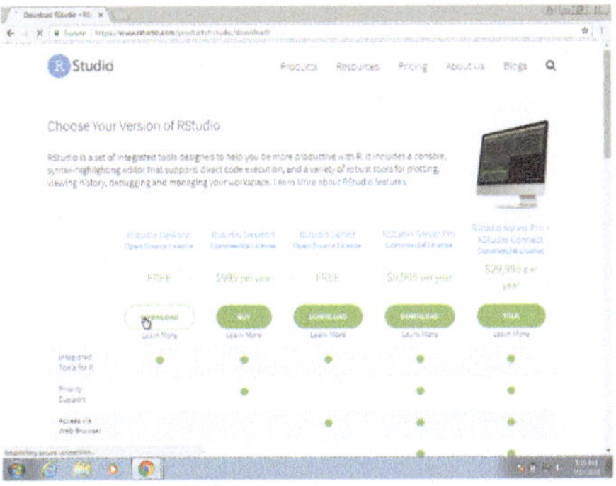

4. Once you select this option, it will take you to a page in which the operating system options are provided. Click the link showing your operating system.

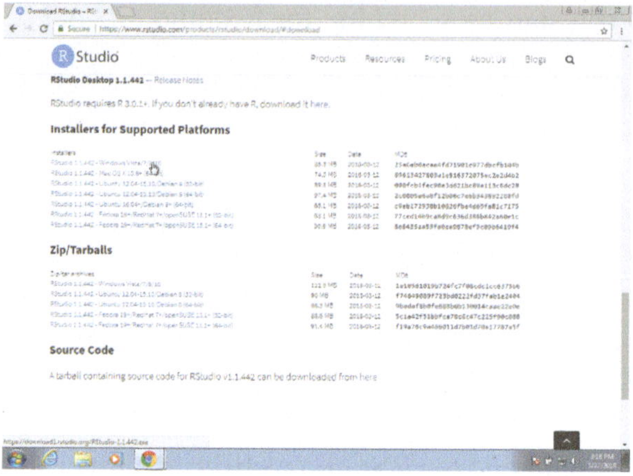

5. Once the installation file is downloaded, click on the downloaded file to start the installation process. It is recommended to click yes on all the defaults.

6. Congratulations! You have installed RStudio.

CHAPTER 2

# GETTING STARTED WITH R AND RSTUDIO

## CONTENTS

R is not a programming language like C or Java. Statisticians developed it as an interactive environment for data analysis. This interactivity is an essential feature in data science because the ability to quickly explore data is a necessity for success in this field.

You can save your work as scripts containing codes that can be executed at any moment. These scripts serve as a record of the analysis you performed, a key feature that facilitates reproducible work. Other attractive features of R are:

- R is free.
- It runs on all major platforms: Windows, Mac Os, UNIX/Linux.
- Scripts and data objects can be shared seamlessly across platforms.
- It is easy for others to contribute add-ons which enables developers to share software implementations of new data science methodologies. This gives R users early access to the latest methods and to tools, which are developed, for a wide variety of disciplines, including ecology, molecular biology, social sciences and geography, just to name a few examples.

## 2.1 THE R CONSOLE

Interactive data analysis usually occurs on the R console that executes commands as you type them. There are several ways to gain access to an R console. One way is to simply start R on your computer. The console looks something like this:

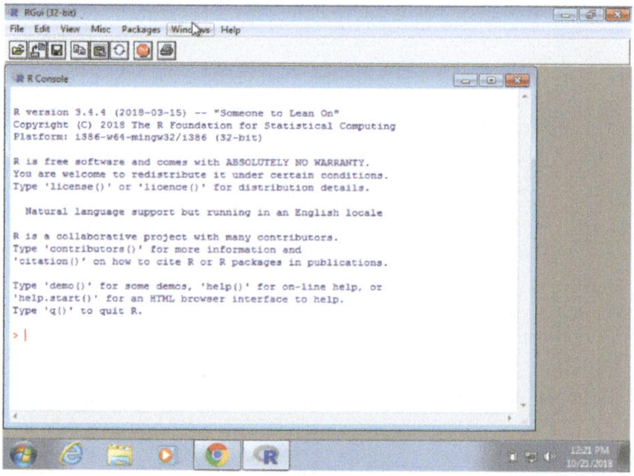


Note that the console starts with greater than sign (>).

As a quick example, try using the console as a calculator:

```
>0.15 * 19.71
[1] 2.96
```

The output is 2.96. [1] is not part of the result but indicates that the result has only the first element which is the number 2.96.

## 2.2 RSTUDIO

When you start RStudio for the first time, you will see three panes. The left pane shows the R console. On the right, the top pane includes three tabs: Environment, History, and Connections, while the bottom pane shows five tabs: File, Plots, Packages, Help, and Viewer. You can click on each tab to move across the different features.

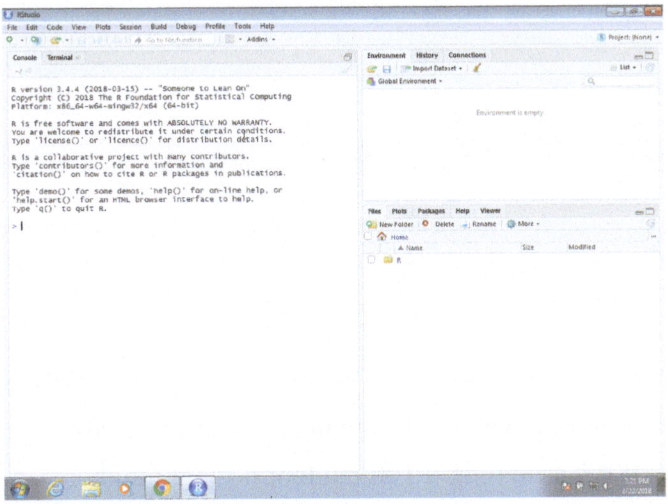


### 2.2.1 Writing Codes

At the R console, we type expressions. The <- symbol (< and -) is the assignment operator. Type the following code and press Enter

```
> x <– 1
```

We assigned the x variable a value of 1. After assigning and pressing Enter, nothing happened. To get the x value, type the following code and press Enter

```
>       print (x)
[1] 1
```

Or simply type the following code:

```
>       x
[1] 1
```

# is the comment character. Anything to the right of #, including # itself, is ignored. It is used to memorize yourself with what this code performs.

Type the following code:

```
>x<–1 #assigning x a value of 1
>x
[1] 1
```

The same result as above.

Note also in the right Environment pane, x is stored and its value is 1.

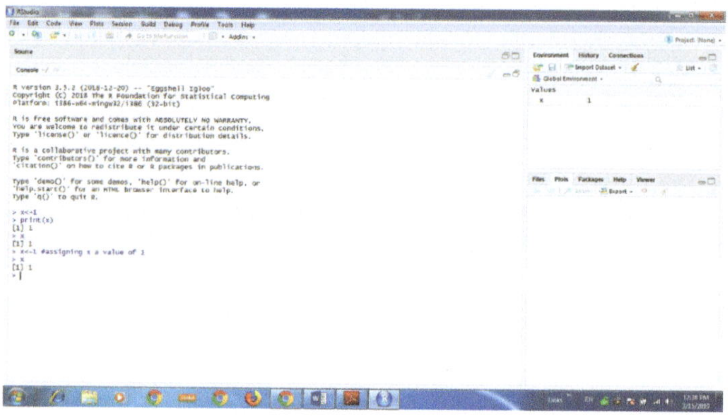

Another example, type the following code in the console:

```
>x<-rnorm(100) # assigning x 100 random normal numbers
>y<-rnorm(100) # assigning y another 100 random normal numbers
>plot (x,y) # a scatter plot of y versus x
```

Note that value of x has changed in the environment in the top right pane. In the bottom right pane, the scatter plot appears. Note the tabs, zoom to enlarge the plot and export to save the plot as .png or .pdf file.

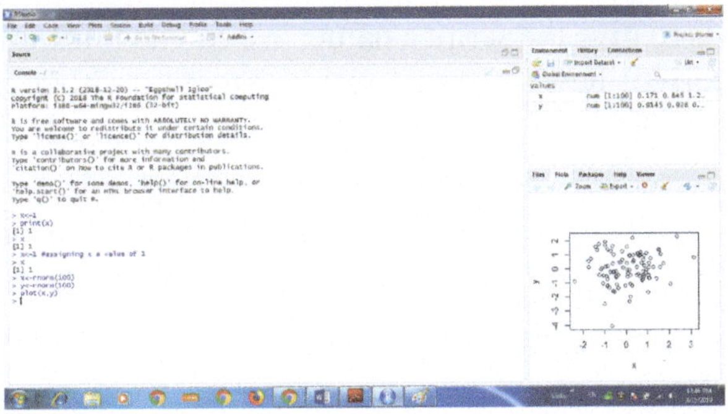

## 2.2.2 Scripts

To start a new script, you can click on File, the New File, then R Script. One of the great advantages of R over point-and-click analysis software is that you can save your work as scripts. You can edit and save these scripts using a text editor as the one in R Studio. In the File tab, click New File and then R Script.

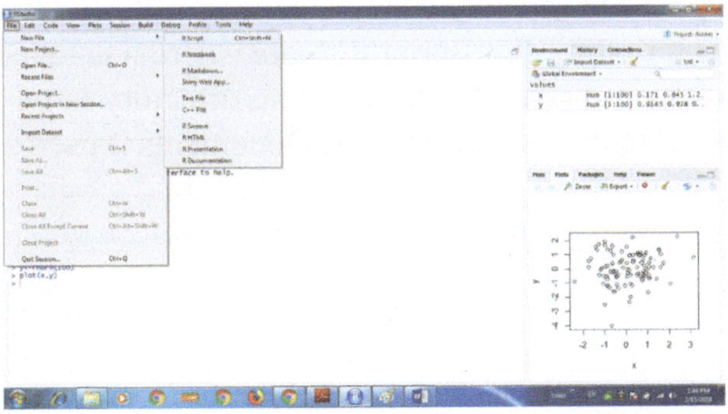

This starts a new pane on the left and it is here where you can start writing your script.

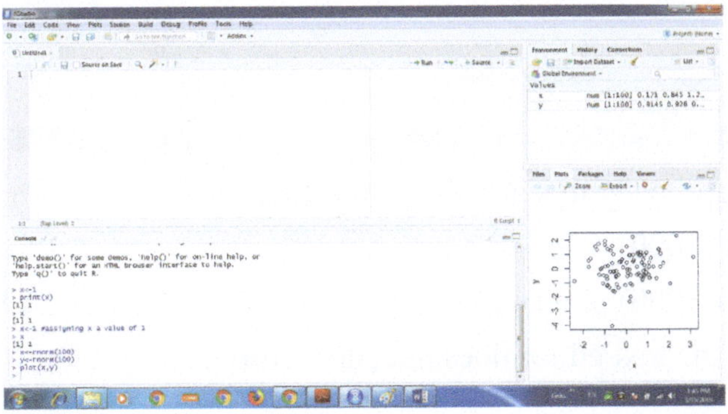

A next step is to give the script a name. We can do this through the editor by saving the current new unnamed script, named untitled, by clicking on the save icon.

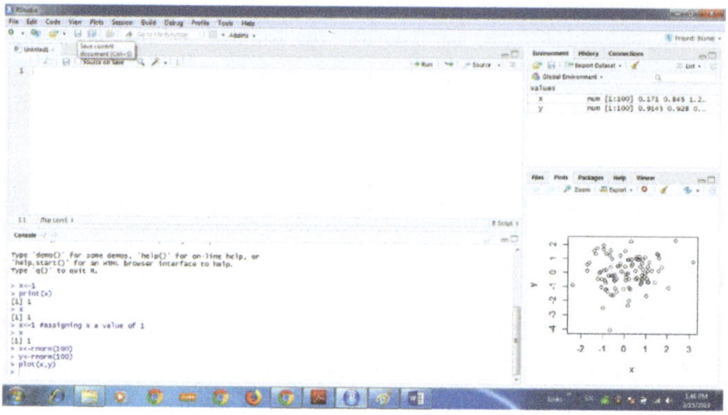

When you ask for the document to be saved for the first time, RStudio will prompt you for a name. You want to use a descriptive name, with no spaces, only hyphens to separate words, and then followed by the suffix .R. We will call this script my-first-script.R. Note where it is saved because that will be your working directory. The working directory is where you place all your files for R. In the example below, Documents is used as the working directory.

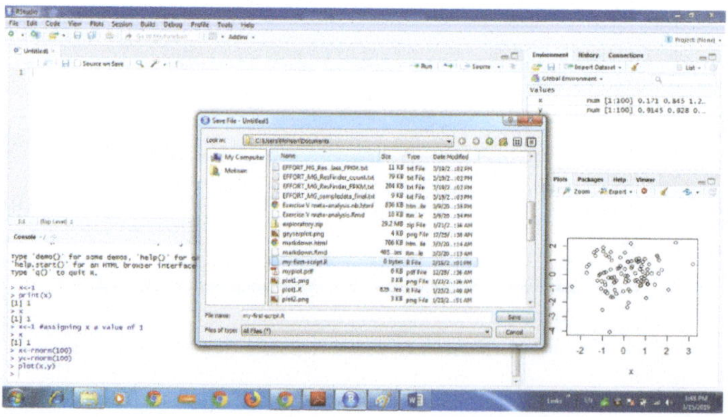

Now we are ready to start editing our first script. Try writing this code:

x<-rnorm (1000)

y<-rnorm (1000) plot (x,y)

Note that the text editor does not start with >.

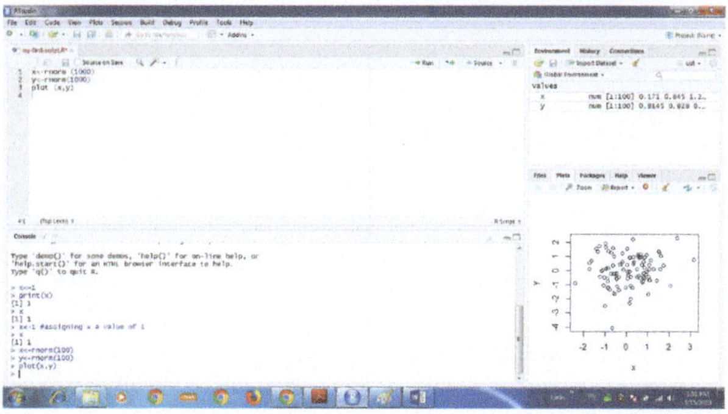

After writing this code, you can try it out by sourcing in the code. To do this, click on the Source button on the upper right side of the editing pane.

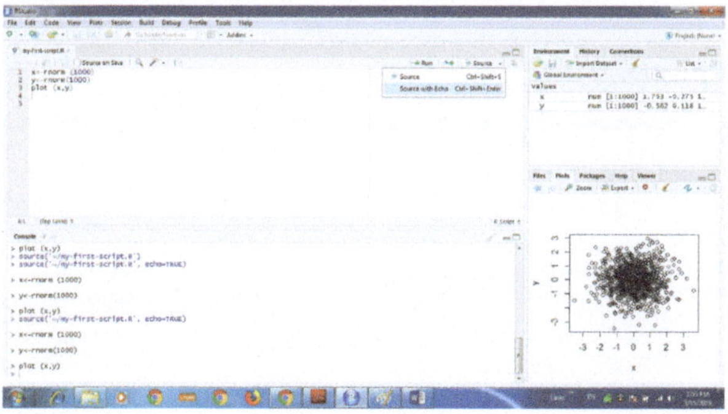

Source has two options, Source will execute the code without printing the code in the console and Source with echo, which will execute the code and printing it to the console. You can try the both options to see the difference as above. In both cases, the plot will appear in the plot pane in the bottom right pane. This plot generates 1000 normal random numbers for x and y. Close the script and open it again from File then open file and choose my-first-script.R.

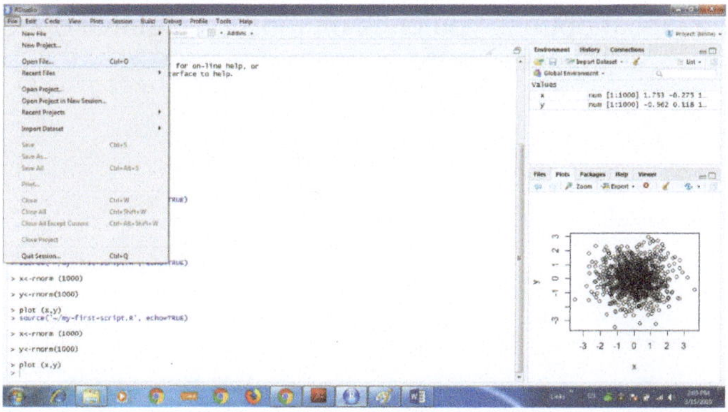

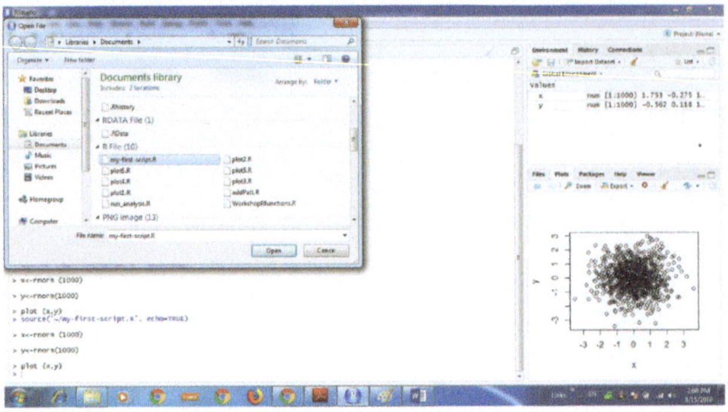

This will open ouR Script again. Modify the code as follows then source it.

```
x<-rnorm(100)
y<-rnorm(100) plot(x,y)
```

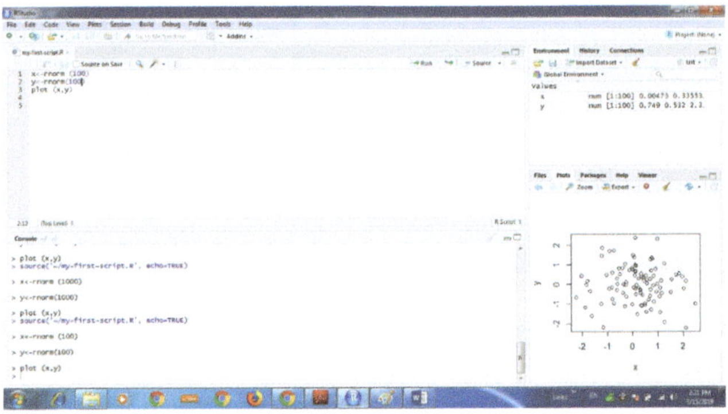

Note how the plot has changed so you can change the script without writing the entire code again.

### 2.2.3 Changing Global Options

To change the global options you click on tools then global options.

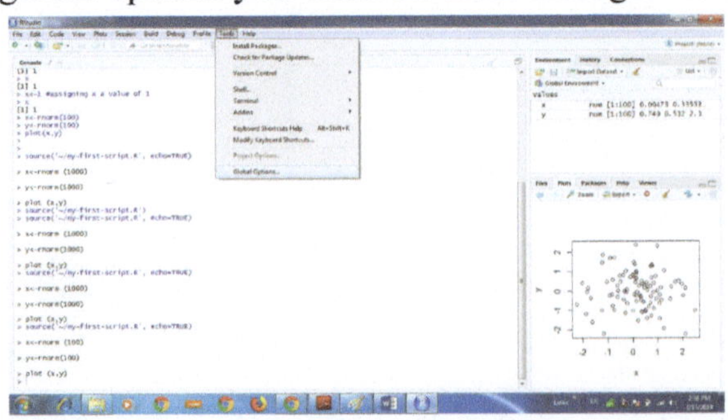

It is recommended to change the Save workspace to .RData on exit to Never and uncheck the Restore. RData into workspace at start. By default, when you exit R saves all the objects you have created into a file called .RData. This is done so that when you restart the session in the same folder, it will load these objects. We find that this causes confusion especially when we share code with another one. To change these options, make your General settings look like this and press Ok.

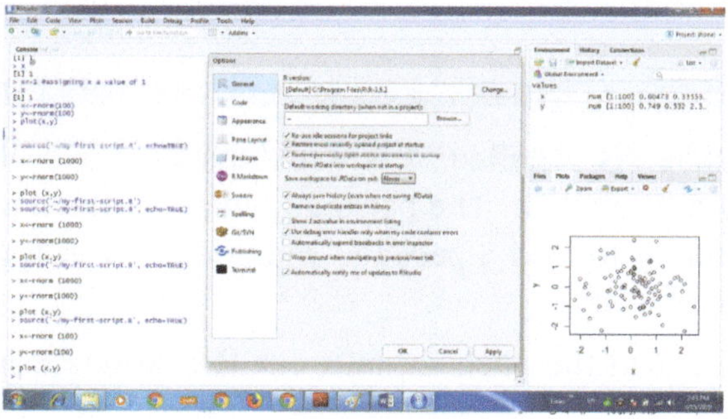

## 2.2.4 Installing R Packages

The functionality provided by a fresh install of R is only a small fraction of what is possible. The extra functionality comes from add-ons (packages) available from developers. There are currently hundreds of these available from CRAN and Bioconductor project and many others shared via other repositories such as GitHub. Anyone can get this extra functionality by installing his needed packages.

In RStudio, you can navigate to the Tools tab and select install packages.

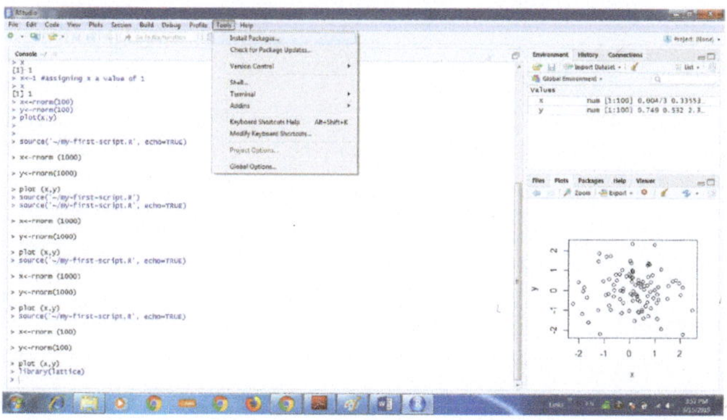

Type tidyverse and dslabs separated by commas and select install.

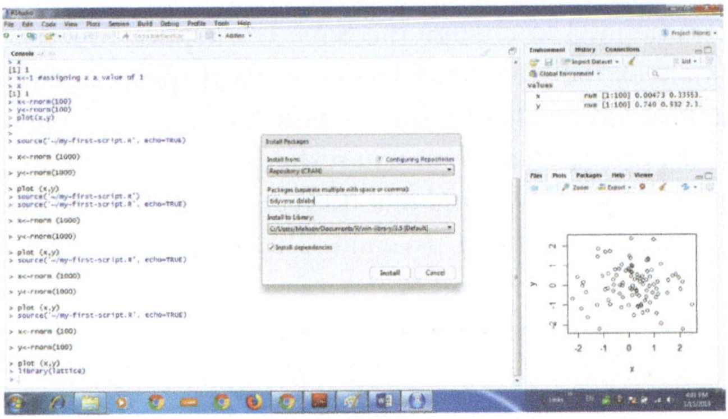

Once a package is installed, it remains installed and only needs to be loaded with library function. The package remains loaded until we quit the R session. If you try to load a package and get an error, it probably means you need to install it first.

We can then load the package into our R sessions using the library() function:

library(dslabs)

library(tidyverse)

You can get some conflicts or errors which is ok now. Note that installing tidyverse actually installs several packages.

Once packages are installed, you can load them into R using library() function. You do not need to install them again, unless you install a fresh version of R. Remember packages are installed in R not RStudio so they must be loaded with library() function for each new session in RStudio.

# CHAPTER 3

# OBJECTS AND FILES

## CONTENTS

## 3.1 WORKING AT R CONSOLE

At the R console (greater than sign ">"), we type expressions or calculations wanted. If used as an interactive calculator, the numeric results are printed.

In the following examples, we will show the different arithmetic operations, addition "+", subtraction "-", division "/", multiplication "*", square root "sqrt", logarithm "log", exponentiation "exp", raising to power "^", and the assignment sign "<-".

Each grey area is a R console area, where we type the command and the results are printed below it.

### 3.1.1 The Addition "+" Sign

Type the following expressions and press Enter

```
100+100
## [1] 200
100+3
## [1] 103
10+3
## [1] 13
10+6
## [1] 16
```

R simply prints the result of each addition operation by default. The [1] shown in the output indicates that the result has its first (and only) element. This element is the result of that operation, and every addition operation has only one number as a result. However, R is a programming language and so it is used not only as a calculator but also to automate some processes or avoid unnecessary repetition.

In many programming environments, the up arrow will cycle through previous commands. Try hitting the up arrow on your keyboard until you get to this command (100 + 100), then change 100 to 1000 and hit Enter. This will save greatly your time.

### 3.1.2 The Subtraction "–" Sign

type the following expressions and press Enter

```
100–90
```

```
## [1] 10
123–25
## [1] 98
102–3
## [1] 99
```

Again, R simply prints the result of each operation with [1] shown before each output.

### 3.1.3 The Division Sign "/"

```
100/3
## [1] 33.33333
6/3
## [1] 2
9/3
## [1] 3
```

### 3.1.4 The Multiplication Sign "*"

```
100*3
## [1] 300
3*2
## [1] 6
3*3
## [1] 9
10*100
## [1] 1000
```

### 3.1.5 The Square Root Sign "sqrt"

sqrt() is a function in R that computes the (principal) square root of the argument which is between its round parentheses.

Anytime you have questions about a particular function, you can access R's built-in help files via the "?" command. For example, if you want more information on the sqrt() function, type ?sqrt without the parentheses that normally follow a function name. Give it a try and see the result at the help tab

**sqrt**(100)

## [1] 10

**sqrt**(9)

## [1] 3

**sqrt**(10)

## [1] 3.162278

### 3.1.6 The Logarithmic Function "log"

log computes logarithms, by default natural logarithms.

log10 computes common (i.e., base 10) logarithms, and log2 computes binary (i.e., base 2) logarithms.

The general form of this function is log(x, base) computes logarithms with base.

x is the first argument and it is the number for which the logarithm is needed.

base is the second argument and it is the base of calculated logarithm.

```
log(100)
## [1] 4.60517
log(100,exp(1))
## [1] 4.60517
log(9)
## [1] 2.197225
log(9,exp(1))
## [1] 2.197225
log(10)
## [1] 2.302585
log(10,exp(1))
## [1] 2.302585
log10(100)
## [1] 2
log(100,exp(10))
```

```
## [1] 0.460517
log2(100)
## [1] 6.643856
log(100,exp(2))
## [1] 2.302585
```

We note that log(number) is the same as log(number,exp(1)), log10(number) is the same as log(number,exp(10)), and log2(number) is the same as log(number,exp(2)). exp is the exponential function and will be discussed below

### 3.1.7 The Exponential Function "exp"

exp computes the exponential function. exp is the inverse function of the log function

```
log(100)
## [1] 4.60517
exp(4.60517)
## [1] 99.99998
exp(log(100))
## [1] 100
```

We note that the result of log(number), when exponentiated, will produce the same number before taking its logarithm

### 3.1.8 The Raising to Power Sign "^"

```
10^3
## [1] 1000
10^–3
## [1] 0.001
10^3.4
## [1] 2511.886
10^–2
## [1] 0.01
```

As you can see, we can raise to positive or negative power or exponents

### 3.1.9 The Assignment Sign “<-”

The “<-” or “=“ symbol is the assignment operator. “<-” is used more commonly. The way you assign a value to a variable in R is by using the assignment operator, which is just a ‘less than’ symbol followed by a ‘minus’ sign.

The # character indicates a comment. Anything to the right of the # (including the # itself) is ignored. This is the only comment character in R. Unlike some other languages, R does not support multi-line comments or comment blocks so you have to put # before every comment line.

```
# nothing printed

x<–3

# autoprinting

x
## [1] 3
# explicit printing

print(x)
## [1] 3
y<–3+5

y
## [1] 8
x+y
## [1] 11
# z<–10

z
## Error in eval(expr, envir, enclos): object ‘z’ not found
```

Think of the assignment operator as an arrow. You are assigning the value on the right side of the arrow to the variable name on the left side of the arrow. In the examples above, x is assigned a value of 3 and y is assigned a value of 8 (3+5)

When y was assigned the result of 5 + 3, R did not print the result of 8. When you use the assignment operator, R assumes that you don't want to see the result immediately, but rather that you intend to use the result for something else later on. To view the contents of the variable x or y, just type x or y and press Enter. You can use the function print() to explicitly print these objects

In the example above, z was assigned a value of 10. IIowever, as the # was typed before that command so it is ignored and when we type z on console, an error is produced

The role of assignment is to use these assigned variables in a second calculation. Instead of retyping 5 + 3 + 3 every time we need it, we can just type x+y.

The [1] shown in the output indicates that x is a vector and 3 is its first (and only) element.

## 3.2 R OBJECTS

R has five basic or "atomic" classes of objects:

- Character;
- Numeric (real numbers);
- Integer;
- Complex;
- Logical (true/false).

Atomic objects mean that they only contain one element. They are also called scalars.

We can see the type of any object with the class function.

### 3.2.1 Character Objects (Strings)

The variable names in R must not start with numbers and must not contain any spaces. Of course, choose your variable names to be meaningful so you can remember them

The character objects are placed within double quotes (strings)

```
# note the error produced

1name <- “Mohsen”

1_name<-”Mohsen”

my_name<-”Mohsen”

my_name

class(my_name)

my_country<-”Egypt”

my_country

class(my_country)
## Error: <text>:4:2: unexpected symbol
## 3:
## 4: 1name
## ^
```

### 3.2.2 Numeric Objects

Numbers in R are generally treated as numeric objects (i.e., double precision real numbers). This means that even if you see a number like “1” or “2” in R, which you might think of as integers, they are likely represented behind the scenes as numeric objects (so something like “1.00” or “2.00”).

If you explicitly want an integer, you need to specify the L suffix. So entering 1 in R gives you a numeric object; entering 1L explicitly gives you an integer object.

```
# numeric object

x<–1
```

```
x
## [1] 1
class(x)
## [1] "numeric"
y<-10

y
## [1] 10
class(y)
## [1] "numeric"
```

Although x and y variables are assigned values of 3 and 8 in previous code chunks, they are now assigned new values which overwrite the old values. Therefore, when x and y are typed at the console, new values are printed out

### 3.2.3 Integer Objects

```
# integer object

x<-1L

x
## [1] 1
class(x)
## [1] "integer"
y<-10L

y
## [1] 10
class(y)
## [1] "integer"
```

### 3.2.4 Logical Objects

```
# logical object

z<-TRUE

l<-FALSE

class(z)
## [1] "logical"
class(l)
## [1] "logical"
z<-T

l<-F

class(z)
## [1] "logical"
class(l)
## [1] "logical"
```

There are only two logical values, TRUE or FALSE. T and F are shorthand values for TRUE and FALSE respectively

### 3.2.5 Complex Objects

Complex objects have imaginary value i

```
# complex object

y<-1i

y
## [1] 0+1i
class(y)
## [1] "complex"
```

There is also a special number Inf which represents infinity. This allows us to represent entities like 1 / 0. This way, Inf can be used in ordinary calculations; e.g., 1 / Inf is 0.

```
1/0
## [1] Inf
```

## 3.3 FILES AND WORKSPACES

In this part, you'll learn how to examine your local workspace in R and begin to explore the relationship between your workspace and the file system of your machine.

Because different operating systems have different conventions with regards to things like file paths, the outputs of these commands may vary across machines. However, it's important to note that R provides a common API (a common set of commands) for interacting with files, that way your code will work across different kinds of computers.

To determine which directory your R session is using as its current working directory, using getwd() function.

**getwd()**

## [1] "E:/R/Anmol project"

**To list all the objects in your local workspace in R, use ls() function.**

**ls()**

## [1] "l" "x" "y" "z"

Assign "John" to variable, my_name2. Now take a look at objects that are in your workspace in R using ls(), you will find that you have an extra object, "my_name2".

```
my_name2<-"John"
```

**ls()**

## [1] "l" "my_name2" "x" "y" "z"

To list all the files in your working directory, use list.files() or dir().

**list.files()**

```
## [1] "~$apter_3_objects (1).docx"
## [2] "~$apter_3_objects.docx"
## [3] "~$apter_4_vectors.docx"
## [4] "~$NDA.docx"
## [5] "chapter_3_objects (1).docx"
## [6] "Chapter_3_objects.docx"
## [7] "Chapter_3_objects.html"
## [8] "Chapter_3_objects.pdf"
## [9] "Chapter_3_objects.Rmd"
## [10] "chapter_4_vectors.docx"
## [11] "chapter_4_vectors.html"
## [12] "chapter_4_vectors.Rmd"
## [13] "Chapter_5_matrices_and_dataframes.docx"
## [14] "Chapter_5_matrices_and_dataframes.html"
## [15] "Chapter_5_matrices_and_dataframes.Rmd"
## [16] "mytest2.R"
## [17] "mytest3.R"
## [18] "NDA.docx"
## [19] "testdir"
## [20] "testdir2"
dir()
## [1] "~$apter_3_objects (1).docx"
## [2] "~$apter_3_objects.docx"
## [3] "~$apter_4_vectors.docx"
## [4] "~$NDA.docx"
## [5] "chapter_3_objects (1).docx"
## [6] "Chapter_3_objects.docx"
## [7] "Chapter_3_objects.html"
## [8] "Chapter_3_objects.pdf"
## [9] "Chapter_3_objects.Rmd"
## [10] "chapter_4_vectors.docx"
```

```
## [11] "chapter_4_vectors.html"
## [12] "chapter_4_vectors.Rmd"
## [13] "Chapter_5_matrices_and_dataframes.docx"
## [14] "Chapter_5_matrices_and_dataframes.html"
## [15] "Chapter_5_matrices_and_dataframes.Rmd"
## [16] "mytest2.R"
## [17] "mytest3.R"
## [18] "NDA.docx"
## [19] "testdir"
## [20] "testdir2"
# to read other data in other folders, provide their full path

dat<-read.csv("E:/R/heart.csv")
```

In the last code, we used read.csv() function to read a comma separated values file (csv file) that is stored in the E compartment in a folder called R.

# CHAPTER 4

# VECTORS AND LISTS

## CONTENTS

## 4.1 NUMERIC VECTORS

A vector can only contain objects of the same class. But of course, like any rule, there is an exception, which is a list. A list is represented as a vector but can contain objects of different classes.

Numeric vectors from its name contain numbers. The easiest way to create a vector is with the c() function, which stands for 'combine'. See the example below.

```
z<-c(1,2,3)


z
## [1] 1 2 3
class(z)
## [1] "numeric"
```

We have used the c() function and stored the result in a variable called z. When we type z on the console to view its contents, there are now no commas separating the values in the output.

Even a single number is considered a vector of length one.

When an R vector is printed, you will notice that an index for the vector is printed in square brackets on the side. The numbers in the square brackets are not part of the vector itself, they are merely part of the printed output. There is a difference between the actual R object and the manner in which that R object is printed to the console. Often, the printed output may have additional parts to make the output more friendly to the users. However, these parts are not inherently part of the object.

You can combine vectors to make a new vector. Create a new vector that contains z, 111, then z again in that order. Assign this vector to a new variable called y and type it on the console, so that you can see its contents.

```
y<-c(z,111,z)


y
## [1] 1 2 3 111 1 2 3
```

### 4.1.1 Vectorized Operations

Numeric vectors can be used in arithmetic expressions. See the following examples:

```
z +10
## [1] 11 12 13
z*10
## [1] 10 20 30
z/3
## [1] 0.3333333 0.6666667 1.0000000
z–10
## [1] –9 –8 –7
z+10*2
## [1] 21 22 23
sqrt(z)
## [1] 1.000000 1.414214 1.732051
log(z)
## [1] 0.0000000 0.6931472 1.0986123
```

This is called vectorized operations. Each operation was applied to each element of z.

Each operation gives you with a vector of the same length as the original vector z.

When given two vectors of the same length, R simply performs the specified arithmetic operation (+, –, *, etc.), element-by-element. If the vectors are of different lengths, R 'recycles' the shorter vector until it is the same length as the longer vector.

When we did z * 10 or z+10 in the above examples, z was a vector of length 3, but 10 is a vector of length 1. Behind the scenes, R is 'recycling' the 10 to make a vector of 10s. In other words, when you ask R to compute z * 10, what it really computes is this: z * c(10, 10, 10).

Examples of vectorized operations when two vectors are of same length.

```
z<-c(1,2,3)
```

```
x<-c(10,20,30)

add<-z+x

add
## [1] 11 22 33
sub<-z-x

sub
## [1] –9 –18 –27
mult<-z*x

mult
## [1] 10 40 90
div<-z/x

div
## [1] 0.1 0.1 0.1
```

If the length of the shorter vector does not divide evenly into the length of the longer vector, R will still apply the 'recycling' method, but will throw a warning to let you know something fishy might be going on.

```
z<-c(1,2,3,4)

x<- c(0,10)

add<-z+x

add
## [1] 1 12 3 14
sub<-z-x
```

```
sub
## [1] 1 –8 3 –6
mult<-z*x

mult
## [1] 0 20 0 40
div<-z/x

div
## [1] Inf 0.2 Inf 0.4
```

Examples of vectorized operations with warning

```
z<- c(1, 2, 3, 4)

x<- c(0, 10, 100)

add<-z+x
## Warning in z + x: longer object length is not a multiple of shorter object
## length
add
## [1] 1 12 103 4
sub<-z-x
## Warning in z - x: longer object length is not a multiple of shorter object
## length
sub
## [1] 1 –8 –97 4
mult<-z*x
## Warning in z * x: longer object length is not a multiple of shorter object
## length
mult
```

```
## [1] 0 20 300 0
div<-z/x
## Warning in z/x: longer object length is not a multiple of shorter object length
div
## [1] Inf 0.20 0.03 Inf
```

### 4.1.2 Sequence of Numbers

1. **colon ":" operator:** It is used to generate regular sequences in the form (from number: to number). The sequences generated are in steps of 1 or –1. See the below examples.

The *from* number need not be smaller than *to* number.

```
1:10
## [1] 1 2 3 4 5 6 7 8 9 10
1:100
## [1] 1 2 3 4 5 6 7 8 9 10 11 12 13 14 15 16 17 18
## [19] 19 20 21 22 23 24 25 26 27 28 29 30 31 32 33 34 35 36
## [37] 37 38 39 40 41 42 43 44 45 46 47 48 49 50 51 52 53 54
## [55] 55 56 57 58 59 60 61 62 63 64 65 66 67 68 69 70 71 72
## [73] 73 74 75 76 77 78 79 80 81 82 83 84 85 86 87 88 89 90
## [91] 91 92 93 94 95 96 97 98 99 100
100:1
## [1] 100 99 98 97 96 95 94 93 92 91 90 89 88 87 86 85 84 83
## [19] 82 81 80 79 78 77 76 75 74 73 72 71 70 69 68 67 66 65
## [37] 64 63 62 61 60 59 58 57 56 55 54 53 52 51 50 49 48 47
## [55] 46 45 44 43 42 41 40 39 38 37 36 35 34 33 32 31 30 29
## [73] 28 27 26 25 24 23 22 21 20 19 18 17 16 15 14 13 12 11
## [91] 10 9 8 7 6 5 4 3 2 1
100:50
## [1] 100 99 98 97 96 95 94 93 92 91 90 89 88 87 86 85 84 83 82
## [20] 81 80 79 78 77 76 75 74 73 72 71 70 69 68 67 66 65 64 63
## [39] 62 61 60 59 58 57 56 55 54 53 52 51 50
```

```
1.5:20
## [1] 1.5 2.5 3.5 4.5 5.5 6.5 7.5 8.5 9.5 10.5 11.5 12.5 13.5 14.5 15.5
## [16] 16.5 17.5 18.5 19.5
1.5:20.5
## [1] 1.5 2.5 3.5 4.5 5.5 6.5 7.5 8.5 9.5 10.5 11.5 12.5 13.5 14.5 15.5
## [16] 16.5 17.5 18.5 19.5 20.5
1.5:20.3
## [1] 1.5 2.5 3.5 4.5 5.5 6.5 7.5 8.5 9.5 10.5 11.5 12.5 13.5 14.5 15.5
## [16] 16.5 17.5 18.5 19.5
–1:–20
## [1] –1 –2 –3 –4 –5 –6 –7 –8 –9 –10 –11 –12 –13 –14 –15 –16 –17 –18 –19
## [20] –20
x<–1:10

class(x)
## [1] “integer”
x<–1.5:20

class (x)
## [1] “numeric”
```

The sequence generated will be of class integer if both from and to numbers are integers and the sequence generated will be of class numeric if from or to number is decimal.

**2. seq() function:** This is same as the colon “:” operator

```
seq(1,10)
## [1] 1 2 3 4 5 6 7 8 9 10
seq(10,1)
## [1] 10 9 8 7 6 5 4 3 2 1
seq(–1,–20)
## [1] –1 –2 –3 –4 –5 –6 –7 –8 –9 –10 –11 –12 –13 –14 –15 –16 –17 –18 –19
## [20] –20
```

```
seq(1.5,20)
## [1] 1.5 2.5 3.5 4.5 5.5 6.5 7.5 8.5 9.5 10.5 11.5 12.5 13.5 14.5 15.5
## [16] 16.5 17.5 18.5 19.5
```

If a single number is supplied to seq() function, a sequence of numbers starting at 1 and ending at that value will be generated

```
seq(100)
## [1] 1 2 3 4 5 6 7 8 9 10 11 12 13 14 15 16 17 18
## [19] 19 20 21 22 23 24 25 26 27 28 29 30 31 32 33 34 35 36
## [37] 37 38 39 40 41 42 43 44 45 46 47 48 49 50 51 52 53 54
## [55] 55 56 57 58 59 60 61 62 63 64 65 66 67 68 69 70 71 72
## [73] 73 74 75 76 77 78 79 80 81 82 83 84 85 86 87 88 89 90
## [91] 91 92 93 94 95 96 97 98 99 100
seq(–100)
## [1] 1 0 –1 –2 –3 –4 –5 –6 –7 –8 –9 –10 –11 –12 –13
## [16] –14 –15 –16 –17 –18 –19 –20 –21 –22 –23 –24 –25 –26 –27 –28
## [31] –29 –30 –31 –32 –33 –34 –35 –36 –37 –38 –39 –40 –41 –42 –43
## [46] –44 –45 –46 –47 –48 –49 –50 –51 –52 –53 –54 –55 –56 –57 –58
## [61] –59 –60 –61 –62 –63 –64 –65 –66 –67 –68 –69 –70 –71 –72 –73
## [76] –74 –75 –76 –77 –78 –79 –80 –81 –82 –83 –84 –85 –86 –87 –88
## [91] –89 –90 –91 –92 –93 –94 –95 –96 –97 –98 –99 –100
```

The by argument of the seq function affords some flexibility for the step increase or decrease. For the step decrease, you must specify the step with minus "-" sign.

```
seq(1,10, by=0.5)
## [1] 1.0 1.5 2.0 2.5 3.0 3.5 4.0 4.5 5.0 5.5 6.0 6.5 7.0 7.5 8.0
## [16] 8.5 9.0 9.5 10.0
seq(1,10,by = 2)
## [1] 1 3 5 7 9
```

```
seq(1,10, by = 3)
## [1] 1 4 7 10
seq(100,50,by = –5)
## [1] 100 95 90 85 80 75 70 65 60 55 50
seq(–100,–50,by = 5)
## [1] –100 –95 –90 –85 –80 –75 –70 –65 –60 –55 –50
```

Another argument, length.out, allows you to specify the desired length of the produced sequence and generate equally spaced values from the from number to number

```
seq(1,100, length.out = 4)
## [1] 1 34 67 100
seq(1,100, length.out = 10)
## [1] 1 12 23 34 45 56 67 78 89 100
seq(1,100, length.out = 15)
## [1] 1.000000 8.071429 15.142857 22.214286 29.285714 36.357143
## [7] 43.428571 50.500000 57.571429 64.642857 71.714286 78.785714
## [13] 85.857143 92.928571 100.000000
seq(1,100, length.out = 20)
## [1] 1.000000 6.210526 11.421053 16.631579 21.842105 27.052632
## [7] 32.263158 37.473684 42.684211 47.894737 53.105263 58.315789
## [13] 63.526316 68.736842 73.947368 79.157895 84.368421 89.578947
## [19] 94.789474 100.000000
```

Another argument, along.with, will generate a sequence that have the same length of that argument length of any vector can be obtained from the length() function.

```
z
## [1] 1 2 3 4
length(z)
```

## [1] 4

**seq**(20,100, along.with = z)

## [1] 20.00000 46.66667 73.33333 100.00000

## *This is the same as*

**seq**(20,100, length.out = 4)

## [1] 20.00000 46.66667 73.33333 100.00000

We noticed that seq() with argument along.with = z is the same as seq() with argument length.out = 4 because the length of z is 4.

3. **rep() function:** It is used to replicate the specified values. rep(10,10) means repeat 10 ten times.

**rep**(10,times=10)

## [1] 10 10 10 10 10 10 10 10 10 10

**rep**(1,times = 11)

## [1] 1 1 1 1 1 1 1 1 1 1 1

**rep**(3,times = 50)

## [1] 3 3 3 3 3 3 3 3 3 3 3 3 3 3 3 3 3 3 3 3 3 3 3 3 3 3 3 3 3 3 3 3 3 3 3 3 3 3

## [39] 3 3 3 3 3 3 3 3 3 3 3 3

rep() function can also use with vectors

**rep**(**c**(1,2,3), times = 10)

## [1] 1 2 3 1 2 3 1 2 3 1 2 3 1 2 3 1 2 3 1 2 3 1 2 3 1 2 3 1 2 3

x<-**c**(1,2,3)

**rep**(x,times = 10)

## [1] 1 2 3 1 2 3 1 2 3 1 2 3 1 2 3 1 2 3 1 2 3 1 2 3 1 2 3 1 2 3

z<–1:3

**rep**(z,times = 10)

## [1] 1 2 3 1 2 3 1 2 3 1 2 3 1 2 3 1 2 3 1 2 3 1 2 3 1 2 3 1 2 3

When vectors are supplied to rep() function, the whole sequence will be repeated.

If you do not want the whole sequence to be repeated, use each argument.

```
rep(c(1,2,3), each = 10)
## [1] 1 1 1 1 1 1 1 1 1 1 2 2 2 2 2 2 2 2 2 2 3 3 3 3 3 3 3 3 3 3
x<-c(1,2,3)

rep(x,each = 10)
## [1] 1 1 1 1 1 1 1 1 1 1 2 2 2 2 2 2 2 2 2 2 3 3 3 3 3 3 3 3 3 3
```

If both times and each are specified, a vector of length = vector length *each* time will be produced.

```
rep(1:2,each = 2, times = 3)
## [1] 1 1 2 2 1 1 2 2 1 1 2 2
rep(c(1,2,3), each = 10, times = 10)
## [1] 1 1 1 1 1 1 1 1 1 1 2 2 2 2 2 2 2 2 2 2 3 3 3 3 3 3 3 3 3 3 1 1 1 1 1 1 1
## [38] 1 1 1 2 2 2 2 2 2 2 2 2 2 3 3 3 3 3 3 3 3 3 3 1 1 1 1 1 1 1 1 1 1 2 2 2 2
## [75] 2 2 2 2 2 2 3 3 3 3 3 3 3 3 3 3 1 1 1 1 1 1 1 1 1 1 2 2 2 2 2 2 2 2 2 2 3
## [112] 3 3 3 3 3 3 3 3 3 1 1 1 1 1 1 1 1 1 1 2 2 2 2 2 2 2 2 2 2 3 3 3 3 3 3 3 3
## [149] 3 3 1 1 1 1 1 1 1 1 1 1 2 2 2 2 2 2 2 2 2 2 3 3 3 3 3 3 3 3 3 3 1 1 1 1 1
## [186] 1 1 1 1 1 2 2 2 2 2 2 2 2 2 2 3 3 3 3 3 3 3 3 3 3 1 1 1 1 1 1 1 1 1 1 2 2
## [223] 2 2 2 2 2 2 2 2 3 3 3 3 3 3 3 3 3 3 1 1 1 1 1 1 1 1 1 1 2 2 2 2 2 2 2 2 2
## [260] 2 3 3 3 3 3 3 3 3 3 3 1 1 1 1 1 1 1 1 1 1 2 2 2 2 2 2 2 2 2 2 3 3 3 3 3 3
## [297] 3 3 3 3
x<-c(1,2,3)

rep(x,each = 10, times = 10)
## [1] 1 1 1 1 1 1 1 1 1 1 2 2 2 2 2 2 2 2 2 2 3 3 3 3 3 3 3 3 3 3 1 1 1 1 1 1 1
## [38] 1 1 1 2 2 2 2 2 2 2 2 2 2 3 3 3 3 3 3 3 3 3 3 1 1 1 1 1 1 1 1 1 1 2 2 2 2
## [75] 2 2 2 2 2 2 3 3 3 3 3 3 3 3 3 3 1 1 1 1 1 1 1 1 1 1 2 2 2 2 2 2 2 2 2 2 3
```

```
## [112] 3 3 3 3 3 3 3 3 3 1 1 1 1 1 1 1 1 1 1 2 2 2 2 2 2 2 2 2 2 3 3 3 3 3 3 3 3
## [149] 3 3 1 1 1 1 1 1 1 1 1 1 2 2 2 2 2 2 2 2 2 2 3 3 3 3 3 3 3 3 3 3 1 1 1 1 1
## [186] 1 1 1 1 1 2 2 2 2 2 2 2 2 2 2 3 3 3 3 3 3 3 3 3 3 1 1 1 1 1 1 1 1 1 1 2 2
## [223] 2 2 2 2 2 2 2 2 3 3 3 3 3 3 3 3 3 3 1 1 1 1 1 1 1 1 1 1 2 2 2 2 2 2 2 2 2
## [260] 2 3 3 3 3 3 3 3 3 3 3 1 1 1 1 1 1 1 1 1 1 2 2 2 2 2 2 2 2 2 2 3 3 3 3 3 3
## [297] 3 3 3 3
```

In the first example, a 12-vector is produced, = 2*2*3 = 12.
In the second and third example, a 300-vector is produced, = 3*10*10 = 300.

## 4.2 INTEGER VECTORS

Integer vectors can be produced by the same functions for numeric vectors except by specifying the "L" suffix.

```
#c() function

z<-c(1L,2L,3L)

z
## [1] 1 2 3
class(z)
## [1] "integer"
# colon operator

z<–1L:10L

z
## [1] 1 2 3 4 5 6 7 8 9 10
class(z)
## [1] "integer"
# seq function
```

```
z<-seq(1L,10L,by=2L)

z
## [1] 1 3 5 7 9
class(z)
## [1] "integer"
# rep function

z<-rep(1L:10L,each=2)

z
## [1] 1 1 2 2 3 3 4 4 5 5 6 6 7 7 8 8 9 9 10 10
class(z)
## [1] "integer"
```

## 4.3 CHARACTER VECTORS

Character vectors are differentiated by having double quotes (““‘). They also can be created with the c() function.

```
my_name<-c("my","name","is","Mohsen")

my_name
## [1] "my"  "name" "is"  "Mohsen"
class(my_name)
## [1] "character"
length(my_name)
## [1] 4
```

We have created a character vector of length = 4

### 4.3.1 paste() Function

paste() function is used to combine vectors after converting them to character vectors.

If applied to a single vector like my_name vector, it will produce a vector of length = 1 if used the collapse argument. The collapse argument will specify the character to separate the results.

```
# This will produce the same vector of length = 4

paste(my_name)
## [1] "my"  "name" "is"  "Mohsen"
# This will produce the vector of length = 1

paste(my_name, collapse = "")
## [1] "mynameisMohsen"
paste(my_name, collapse = " ")
## [1] "my name is Mohsen"
paste(my_name, collapse = "-")
## [1] "my-name-is-Mohsen"
paste(my_name, collapse = ",")
## [1] "my,name,is,Mohsen"
```

If applied to different vectors, they will be combined term by term to produce a single vector. The terms are separated by the sep argument. Vector arguments are recycled as necessary. The vectors are combined with the same order as they are applied to paste() function.

```
my_name
## [1] "my"  "name" "is"  "Mohsen"
paste(my_name, 1:4, sep="")
## [1] "my1"  "name2" "is3"  "Mohsen4"
paste(my_name, 1:4, sep=" ")
```

```
## [1] "my 1" "name 2" "is 3" "Mohsen 4"
paste(my_name, 1:4, sep=",")
## [1] "my,1" "name,2" "is,3" "Mohsen,4"
paste(my_name, 1:3, sep=",")
## [1] "my,1" "name,2" "is,3" "Mohsen,1"
paste(my_name, 1:2, sep=",")
## [1] "my,1" "name,2" "is,1" "Mohsen,2"
paste(my_name, 5, sep="_")
## [1] "my_5" "name_5" "is_5" "Mohsen_5"
paste( 1:2, my_name, sep=",")
## [1] "1,my" "2,name" "1,is" "2,Mohsen"
```

An example, the state.name, state.abb, state.region are built in character vectors in R that contain the state names, abbreviations, and regions of each state. Each vector has a length of only 50 since these data were collected in 1970s.

```
state.name
## [1] "Alabama" "Alaska" "Arizona" "Arkansas"
## [5] "California" "Colorado" "Connecticut" "Delaware"
## [9] "Florida" "Georgia" "Hawaii" "Idaho"
## [13] "Illinois" "Indiana" "Iowa" "Kansas"
## [17] "Kentucky" "Louisiana" "Maine" "Maryland"
## [21] "Massachusetts" "Michigan" "Minnesota" "Mississippi"
## [25] "Missouri" "Montana" "Nebraska" "Nevada"
## [29] "New Hampshire" "New Jersey" "New Mexico" "New York"
## [33] "North Carolina" "North Dakota" "Ohio" "Oklahoma"
## [37] "Oregon" "Pennsylvania" "Rhode Island" "South Carolina"
## [41] "South Dakota" "Tennessee" "Texas" "Utah"
## [45] "Vermont" "Virginia" "Washington" "West Virginia"
## [49] "Wisconsin" "Wyoming"
state.abb
```

## [1] "AL" "AK" "AZ" "AR" "CA" "CO" "CT" "DE" "FL" "GA" "HI" "ID" "IL" "IN" "IA"

## [16] "KS" "KY" "LA" "ME" "MD" "MA" "MI" "MN" "MS" "MO" "MT" "NE" "NV" "NH" "NJ"

## [31] "NM" "NY" "NC" "ND" "OH" "OK" "OR" "PA" "RI" "SC" "SD" "TN" "TX" "UT" "VT"

## [46] "VA" "WA" "WV" "WI" "WY"

state.region

## [1] South West West South West

## [6] West Northeast South South South

## [11] West West North Central North Central North Central

## [16] North Central South South Northeast South

## [21] Northeast North Central North Central South North Central

## [26] West North Central West Northeast Northeast

## [31] West Northeast South North Central North Central

## [36] South West Northeast Northeast South

## [41] North Central South South West Northeast

## [46] South West South North Central West

## Levels: Northeast South North Central West

We can paste these vectors together to produce a single vector of length = 50 but contain the state name, abb, and region in each element.

**paste**(state.name, state.abb, state.region, sep = "-")

## [1] "Alabama-AL-South" "Alaska-AK-West"

## [3] "Arizona-AZ-West" "Arkansas-AR-South"

## [5] "California-CA-West" "Colorado-CO-West"

## [7] "Connecticut-CT-Northeast" "Delaware-DE-South"

## [9] "Florida-FL-South" "Georgia-GA-South"

## [11] "Hawaii-HI-West" "Idaho-ID-West"

## [13] "Illinois-IL-North Central" "Indiana-IN-North Central"

## [15] "Iowa-IA-North Central" "Kansas-KS-North Central"

```
## [17] "Kentucky-KY-South"   "Louisiana-LA-South"
## [19] "Maine-ME-Northeast"   "Maryland-MD-South"
## [21] "Massachusetts-MA-Northeast" "Michigan-MI-North Central"
## [23] "Minnesota-MN-North Central" "Mississippi-MS-South"
## [25] "Missouri-MO-North Central"  "Montana-MT-West"
## [27] "Nebraska-NE-North Central"  "Nevada-NV-West"
## [29] "New Hampshire-NH-Northeast" "New Jersey-NJ-Northeast"
## [31] "New Mexico-NM-West"   "New York-NY-Northeast"
## [33] "North Carolina-NC-South"  "North Dakota-ND-North Central"
## [35] "Ohio-OH-North Central"   "Oklahoma-OK-South"
## [37] "Oregon-OR-West"    "Pennsylvania-PA-Northeast"
## [39] "Rhode Island-RI-Northeast"  "South Carolina-SC-South"
## [41] "South Dakota-SD-North Central" "Tennessee-TN-South"
## [43] "Texas-TX-South"    "Utah-UT-West"
## [45] "Vermont-VT-Northeast"   "Virginia-VA-South"
## [47] "Washington-WA-West"   "West Virginia-WV-South"
## [49] "Wisconsin-WI-North Central" "Wyoming-WY-West"
length(state.abb)
## [1] 50
length(state.name)
## [1] 50
length(state.region)
## [1] 50
x<-paste(state.name, state.abb, state.region, sep = "-")

length(x)
## [1] 50
```

## 4.4 LOGICAL VECTORS

Logical vectors contain only two values, TRUE, and FALSE. Logical vectors are commonly produced by using logical operators:

- greater than “>”
- smaller than “<”
- greater than or equal “>=”
- smaller than or equal “<=”
- equal “==”
- nonequal “!=”

```
x<–1:10

x >3
## [1] FALSE FALSE FALSE TRUE TRUE TRUE TRUE TRUE TRUE
TRUE
x < 3
## [1] TRUE TRUE FALSE FALSE FALSE FALSE FALSE FALSE FALSE
FALSE
x >= 1
## [1] TRUE TRUE TRUE TRUE TRUE TRUE TRUE TRUE TRUE TRUE
x <=5
## [1] TRUE TRUE TRUE TRUE TRUE FALSE FALSE FALSE FALSE
FALSE
x == 5
## [1] FALSE FALSE FALSE FALSE TRUE FALSE FALSE FALSE
FALSE FALSE
x !=5
## [1] TRUE TRUE TRUE TRUE FALSE TRUE TRUE TRUE TRUE
TRUE
```

In each operation, TRUE or FALSE is produced according to if the condition is true or not.

R treats TRUE as ones and FALSE as zeros, so by using logical operators and sum() or mean() function, we can explore long vectors with simple codes.

```
x<–1:1000
```

```
sum(x > 300)
## [1] 700
mean(x >300)
## [1] 0.7
```

x is a vector with values from 1 to 1000 in 1 steps so it has 1000 elements (numbers).

By typing sum(x>300), we get 700. This means that there are 700 elements (numbers) in the x vector that are larger than 300.

By typing mean(x>300), we get 0.7. This means that the proportion of elements (numbers) in the x vector that are larger than 300 is 0.7 (=700/1000) or 70%.

```
x<-seq(0,100, by = 5)

sum(x >= 30)
## [1] 15
mean(x >= 30)
## [1] 0.7142857
sum(x == 55)
## [1] 1
mean(x == 55)
## [1] 0.04761905
sum(x != 30)
## [1] 20
mean(x != 30)
## [1] 0.952381
sum(x >= 30 & x != 55)
## [1] 14

mean(x >= 30 & x !=55)
```

```
## [1] 0.6666667
sum(x >= 30 | x != 55)
## [1] 21
mean(x >= 30 | x !=55)
## [1] 1
```

x is a vector with values from 0 to 100 in 5 steps so it has 21 elements (numbers).

By typing sum(x>=30), we get 15. This means that there are 15 elements (numbers) in the x vector that are larger than or equal to 30.

By typing mean(x>=30), we get 0.714. This means that the proportion of elements (numbers) in the x vector that are larger than or equal to 30 is 0.714(=15/21) or 71.4%

By typing sum(x ==55), we get 1. This means that there is only 1 element (number) in the x vector that is equal to 55.

By typing mean(x ==55), we get 0.0476. This means that the proportion of elements (numbers) in the x vector that are equal to 55 is 0.0476 or 4.76%

By typing sum(x !=30), we get 20. This means that there are 20 elements (numbers) in the x vector that are not equal to 30.

By typing mean(x !=30), we get 0.952. This means that the proportion of elements (numbers) in the x vector that are not equal to 30 is 0.952 or 95.2%

By typing sum(x >= 30 & x != 55), we get 14. This means that there are 14 elements (numbers) in the x vector that are not equal to 55 and larger than or equal to 30.

By typing mean(x >= 30 & x != 55), we get 0.67. This means that the proportion of elements (numbers) in the x vector that are not equal to 55 and larger than or equal to 30 is 0.67 or 67%.

By typing sum(x >= 30 | x !=55), we get 21. This means that all the elements (numbers) in the x vector are either not equal to 55 or larger than or equal to 30.

By typing mean(x >= 30 | x !=55), we get 1. This means that the proportion of elements (numbers) in the x vector that are either not equal to 55 or larger than or equal to 30 is 1 or 100%.

### 4.4.1 All() and Any() Functions

Two important functions related to logical vectors are the all() function and any() function:

all() function, when applied to a logical vector, will yield TRUE if all elements of the logical vector are TRUE and FALSE otherwise

any() function, when applied to a logical vector, will yield TRUE if any element of the logical vector is TRUE and FALSE otherwise

```
x<-10:100

x
## [1] 10 11 12 13 14 15 16 17 18 19 20 21 22 23 24 25 26 27 28
## [20] 29 30 31 32 33 34 35 36 37 38 39 40 41 42 43 44 45 46 47
## [39] 48 49 50 51 52 53 54 55 56 57 58 59 60 61 62 63 64 65 66
## [58] 67 68 69 70 71 72 73 74 75 76 77 78 79 80 81 82 83 84 85
## [77] 86 87 88 89 90 91 92 93 94 95 96 97 98 99 100
all(x==10)
## [1] FALSE
any(x==10)
## [1] TRUE
all(x>=10)
## [1] TRUE
any(x>=10)
## [1] TRUE
```

## 4.5 COMPLEX VECTORS

Complex vectors contain complex objects (numbers) that have imaginary part added to it. Complex vectors and complex objects are rarely used in data analysis.

```
## complex numbers
```

```
x<–1i

class(x)
## [1] "complex"
y<–2i

class(y)
## [1] "complex"
## complex vector

z<-c(1i,2i)

class(z)
## [1] "complex"
```

## 4.6 IMPLICIT COERCION

When different classes of R objects get mixed together in a vector, but vectors must have objects of the same type, so coercion occurs so that every element in the vector is of the same class.

R tries to do this by finding a way to represent all of the objects in the vector in a reasonable fashion.

Combining a numeric object with a character object will create a character vector, where numbers are represented as strings in this case.

Combining a numeric object with a logical object will create a numeric vector, where TRUE is represented as 1 and FALSE is represented as 0 in this case.

Combining a numeric object with an integer object will create a numeric vector. Combining a character object with a logical object will create a character vector, where TRUE and FALSE are represented as strings in this case.

```
# numeric and character
```

```
x<-c(1,2,"a")

x
## [1] "1" "2" "a"
class(x)
## [1] "character"
# numeric and logical

x<-c(1,2,TRUE)

x
## [1] 1 2 1
class(x)
## [1] "numeric"
x<-c(1,2,FALSE)

x
## [1] 1 2 0
class(x)
## [1] "numeric"
# numeric and integer

x<-c(1,2,10L)

x
## [1] 1 2 10
class(x)
## [1] "numeric"
# character and logical

x<-c("a","b",TRUE, FALSE)
```

```
x
## [1] "a" "b" "TRUE" "FALSE"
class(x)
## [1] "character"
```

## 4.7 EXPLICIT COERCION

Vectors can be explicitly coerced from one class to another using the as.* functions.

as.numeric, as.integer, as.character, as.logical are examples of the as.* family that will convert the vector to its class.

```
x<-1:10

class(x)
## [1] "integer"
as.numeric(x)
## [1] 1 2 3 4 5 6 7 8 9 10
as.character(x)
## [1] "1" "2" "3" "4" "5" "6" "7" "8" "9" "10"
as.logical(x)
## [1] TRUE TRUE TRUE TRUE TRUE TRUE TRUE TRUE TRUE TRUE
```

Note that R has converted 1 and all larger numbers to TRUE when converted to a logical vector.

When R cannot figure out how to coerce a vector, NAs (missing values) are produced with a warning message.

```
x<-c("a","b","c")

class(x)
## [1] "character"
```

```
as.numeric(x)
## Warning: NAs introduced by coercion
## [1] NA NA NA
as.integer(x)
## Warning: NAs introduced by coercion
## [1] NA NA NA
as.logical(x)
## [1] NA NA NA
```

## 4.8 LISTS

Lists are special type of vectors because:

1. Lists can contain objects of different classes.
2. Lists can group individual values as well as whole R vectors. The grouped vectors need not be of the same length.

lists can be created using the list() function and list() function creates a list the same way c() creates a vector. Separate each element in the list with a comma.

### 4.8.1 Lists of Individual Elements

Example of a list with individual values.

```
x<-list(10,"a",TRUE)

x
## [[1]]
## [1] 10
##
## [[2]]
## [1] "a"
##
## [[3]]
## [1] TRUE
```

The first element is numeric (10), the second element is a character ("a"), and the third element is a logical value.

Note that each element of the list begins with double square brackets. Lists can have names for its elements.

```
x<-list(number = 10, character = "a", logical = TRUE)

x
## $number
## [1] 10
##
## $character
## [1] "a"
##
## $logical
## [1] TRUE
```

Note that each element of the list now begins with the element name instead of double square brackets.

### 4.8.2 Lists of Whole Vectors

Example of a list with whole vectors.

```
x<-list(state_names = state.name, numeric_vector = 1:100, logical_vector =
(seq(1,50,10) > 10))

x
## $state_names
## [1] "Alabama" "Alaska" "Arizona" "Arkansas"
## [5] "California" "Colorado" "Connecticut" "Delaware"
## [9] "Florida" "Georgia" "Hawaii" "Idaho"
## [13] "Illinois" "Indiana" "Iowa" "Kansas"
```

```
## [17] "Kentucky" "Louisiana" "Maine" "Maryland"
## [21] "Massachusetts" "Michigan" "Minnesota" "Mississippi"
## [25] "Missouri" "Montana" "Nebraska" "Nevada"
## [29] "New Hampshire" "New Jersey" "New Mexico" "New York"
## [33] "North Carolina" "North Dakota" "Ohio" "Oklahoma"
## [37] "Oregon" "Pennsylvania" "Rhode Island" "South Carolina"
## [41] "South Dakota" "Tennessee" "Texas" "Utah"
## [45] "Vermont" "Virginia" "Washington" "West Virginia"
## [49] "Wisconsin" "Wyoming"
##
## $numeric_vector
## [1] 1 2 3 4 5 6 7 8 9 10 11 12 13 14 15 16 17 18
## [19] 19 20 21 22 23 24 25 26 27 28 29 30 31 32 33 34 35 36
## [37] 37 38 39 40 41 42 43 44 45 46 47 48 49 50 51 52 53 54
## [55] 55 56 57 58 59 60 61 62 63 64 65 66 67 68 69 70 71 72
## [73] 73 74 75 76 77 78 79 80 81 82 83 84 85 86 87 88 89 90
## [91] 91 92 93 94 95 96 97 98 99 100
##
## $logical_vector
## [1] FALSE TRUE TRUE TRUE TRUE
```

The first element of the list is a character vector of length 50 that contains the 50 state names.

The second element of the list is a numeric vector of length 100 that contains numbers from 1 to 100.

The third element of the list is a logical vector of length 5 that contains 1 FALSE and 4 TRUE. This is testing if the generated sequence, seq(1,50,10), is greater than 10 or not.

The names() is used to get the element names or change the element names of a list.

```
x<-list(month_names = month.name, numeric_vector = 1:12, character_
```

```
vector = letters)

x
## $month_names
## [1] "January" "February" "March"  "April"  "May"  "June"
## [7] "July"  "August" "September" "October" "November" "December"
##
## $numeric_vector
## [1] 1 2 3 4 5 6 7 8 9 10 11 12
##
## $character_vector
## [1] "a" "b" "c" "d" "e" "f" "g" "h" "i" "j" "k" "l" "m" "n" "o" "p" "q"
"r" "s"
## [20] "t" "u" "v" "w" "x" "y" "z"
names(x)
## [1] "month_names"  "numeric_vector" "character_vector"
names(x)<-c("months","vector1","vector2")

x
## $months
## [1] "January" "February" "March"  "April"  "May"  "June"
## [7] "July"  "August" "September" "October" "November" "December"
##
## $vector1
## [1] 1 2 3 4 5 6 7 8 9 10 11 12
##
## $vector2
## [1] "a" "b" "c" "d" "e" "f" "g" "h" "i" "j" "k" "l" "m" "n" "o" "p" "q"
"r" "s"
## [20] "t" "u" "v" "w" "x" "y" "z"
```

# CHAPTER 5

# MATRICES AND DATAFRAMES

## CONTENTS

## 5.1 BUILDING MATRICES WITH MATRIX() FUNCTION

Matrices are tabular data vectors in R. Tabular data means it consists of rows and columns.

Unlike dataframes, matrices can only contain 1 type of R objects (numeric, character, logical, …)

Matrices can be created using the matrix() function. They are constructed column-wise so it starts in the upper left corner and ends in the bottom right corner.

The desired number of rows are specified by nrow argument and the desired number of columns are specified by ncol argument. Of course, nrow*ncol = matrix length.

```
x<- matrix(1:10, nrow = 5, ncol = 2)

x
## [,1] [,2]
## [1,] 1 6
## [2,] 2 7
## [3,] 3 8
## [4,] 4 9
## [5,] 5 10
x<- matrix(1:10, nrow = 2, ncol = 5)

x
## [,1] [,2] [,3] [,4] [,5]
## [1,] 1 3 5 7 9
## [2,] 2 4 6 8 10
y<-matrix(11:25, nrow = 3, ncol = 5)

y
## [,1] [,2] [,3] [,4] [,5]
## [1,] 11 14 17 20 23
## [2,] 12 15 18 21 24
```

```
## [3,] 13 16 19 22 25
y<-matrix(11:25, nrow = 5, ncol = 3)

y
##  [,1] [,2] [,3]
## [1,] 11 16 21
## [2,] 12 17 22
## [3,] 13 18 23
## [4,] 14 19 24
## [5,] 15 20 25
# matrix of character vectors

z<-matrix(state.name, nrow = 10,ncol = 5)

z
##  [,1]  [,2]  [,3]  [,4]
## [1,] "Alabama"  "Hawaii" "Massachusetts" "New Mexico"
## [2,] "Alaska"  "Idaho"  "Michigan"  "New York"
## [3,] "Arizona"  "Illinois" "Minnesota"  "North Carolina"
## [4,] "Arkansas" "Indiana" "Mississippi" "North Dakota"
## [5,] "California" "Iowa"  "Missouri"  "Ohio"
## [6,] "Colorado" "Kansas" "Montana"  "Oklahoma"
## [7,] "Connecticut" "Kentucky" "Nebraska"  "Oregon"
## [8,] "Delaware" "Louisiana" "Nevada"  "Pennsylvania"
## [9,] "Florida"  "Maine"  "New Hampshire" "Rhode Island"
## [10,] "Georgia"  "Maryland" "New Jersey" "South Carolina"
##  [,5]
## [1,] "South Dakota"
## [2,] "Tennessee"
## [3,] "Texas"
```

```
## [4,] "Utah"
## [5,] "Vermont"
## [6,] "Virginia"
## [7,] "Washington"
## [8,] "West Virginia"
## [9,] "Wisconsin"
## [10,] "Wyoming"
z<-matrix(state.name, nrow = 2,ncol = 25)

z
## [,1] [,2] [,3] [,4] [,5] [,6]
## [1,] "Alabama" "Arizona" "California" "Connecticut" "Florida" "Hawaii"
## [2,] "Alaska" "Arkansas" "Colorado" "Delaware" "Georgia" "Idaho"
## [,7] [,8] [,9] [,10] [,11] [,12]
## [1,] "Illinois" "Iowa" "Kentucky" "Maine" "Massachusetts" "Minnesota"
## [2,] "Indiana" "Kansas" "Louisiana" "Maryland" "Michigan" "Mississippi"
## [,13] [,14] [,15] [,16] [,17]
## [1,] "Missouri" "Nebraska" "New Hampshire" "New Mexico" "North Carolina"
## [2,] "Montana" "Nevada" "New Jersey" "New York" "North Dakota"
## [,18] [,19] [,20] [,21] [,22]
## [1,] "Ohio" "Oregon" "Rhode Island" "South Dakota" "Texas"
## [2,] "Oklahoma" "Pennsylvania" "South Carolina" "Tennessee" "Utah"
## [,23] [,24] [,25]
## [1,] "Vermont" "Washington" "Wisconsin"
## [2,] "Virginia" "West Virginia" "Wyoming"
z<-matrix(state.name, nrow = 25,ncol = 2)

z
## [,1] [,2]
```

```
## [1,] "Alabama" "Montana"
## [2,] "Alaska" "Nebraska"
## [3,] "Arizona" "Nevada"
## [4,] "Arkansas" "New Hampshire"
## [5,] "California" "New Jersey"
## [6,] "Colorado" "New Mexico"
## [7,] "Connecticut" "New York"
## [8,] "Delaware" "North Carolina"
## [9,] "Florida" "North Dakota"
## [10,] "Georgia" "Ohio"
## [11,] "Hawaii" "Oklahoma"
## [12,] "Idaho" "Oregon"
## [13,] "Illinois" "Pennsylvania"
## [14,] "Indiana" "Rhode Island"
## [15,] "Iowa" "South Carolina"
## [16,] "Kansas" "South Dakota"
## [17,] "Kentucky" "Tennessee"
## [18,] "Louisiana" "Texas"
## [19,] "Maine" "Utah"
## [20,] "Maryland" "Vermont"
## [21,] "Massachusetts" "Virginia"
## [22,] "Michigan" "Washington"
## [23,] "Minnesota" "West Virginia"
## [24,] "Mississippi" "Wisconsin"
## [25,] "Missouri" "Wyoming"
```

Data vector supplied to matrix() function has a length of 10, 15, or 50 so nrow*ncol = 10, 15, or 50.

In each case, the matrix was filled column-wise.

Note that rows are indexed with square brackets with the row number before the comma ([1,],[2,],[3,],…), while the columns are indexed with square brackets with the column number after the comma ([,1],[,2],[,3], …).

If nrow*ncol longer than the vector length, recycling will occur with a warning.

```
x<- matrix(1:10, nrow = 2, ncol = 6)
## Warning in matrix(1:10, nrow = 2, ncol = 6): data length [10] is not a sub-
## multiple or multiple of the number of columns [6]
x
## [,1] [,2] [,3] [,4] [,5] [,6]
## [1,] 1 3 5 7 9 1
## [2,] 2 4 6 8 10 2
y<-matrix(11:25, nrow = 3, ncol = 10)

y
## [,1] [,2] [,3] [,4] [,5] [,6] [,7] [,8] [,9] [,10]
## [1,] 11 14 17 20 23 11 14 17 20 23
## [2,] 12 15 18 21 24 12 15 18 21 24
## [3,] 13 16 19 22 25 13 16 19 22 25
y<-matrix(11:25, nrow = 3, ncol = 7)
## Warning in matrix(11:25, nrow = 3, ncol = 7): data length [15] is not a
sub-
## multiple or multiple of the number of columns [7]
y
## [,1] [,2] [,3] [,4] [,5] [,6] [,7]
## [1,] 11 14 17 20 23 11 14
## [2,] 12 15 18 21 24 12 15
## [3,] 13 16 19 22 25 13 16
```

Note the warning message produced in each case.

If nrow*ncol shorter than the vector length, matrix building will stop at the specified rows and columns with a warning.

```
y<-matrix(1:24, nrow = 3, ncol = 7)
```

```
## Warning in matrix(1:24, nrow = 3, ncol = 7): data length [24] is not a sub-
## multiple or multiple of the number of columns [7]
y
## [,1] [,2] [,3] [,4] [,5] [,6] [,7]
## [1,] 1 4 7 10 13 16 19
## [2,] 2 5 8 11 14 17 20
## [3,] 3 6 9 12 15 18 21
y<-matrix(1:24, nrow = 2, ncol = 7)
## Warning in matrix(1:24, nrow = 2, ncol = 7): data length [24] is not a sub-
## multiple or multiple of the number of columns [7]
y
## [,1] [,2] [,3] [,4] [,5] [,6] [,7]
## [1,] 1 3 5 7 9 11 13
## [2,] 2 4 6 8 10 12 14
y<-matrix(1:24, nrow = 2, ncol = 5)
## Warning in matrix(1:24, nrow = 2, ncol = 5): data length [24] is not a sub-
## multiple or multiple of the number of columns [5]
y
## [,1] [,2] [,3] [,4] [,5]
## [1,] 1 3 5 7 9
## [2,] 2 4 6 8 10
```

Note the warning message produced in each case.

If you want to fill the constructed matrix row-wise, use the byrow = TRUE argument.

```
y<-matrix(1:24, nrow = 3, ncol = 8)

y
## [,1] [,2] [,3] [,4] [,5] [,6] [,7] [,8]
## [1,] 1 4 7 10 13 16 19 22
```

```
## [2,] 2 5 8 11 14 17 20 23
## [3,] 3 6 9 12 15 18 21 24
y<-matrix(1:24, nrow = 3, ncol = 8, byrow = TRUE)

y
## [,1] [,2] [,3] [,4] [,5] [,6] [,7] [,8]
## [1,] 1 2 3 4 5 6 7 8
## [2,] 9 10 11 12 13 14 15 16
## [3,] 17 18 19 20 21 22 23 24
```

Note the difference between the two constructed matrices. The first matrix is constructed column-wise, while the second matrix was constructed row_ wise.

### 5.1.1 Dimnames

You can specify the row names and column names by using the dimnames argument. dimnames takes a list of two vectors, the first is row names and second is column names.

```
y<-matrix(1:24, nrow = 3, ncol = 8,

  dimnames = list(c("row1","row2","row3"),

    c("column1","column2","column3",
    "column4","column5","column6",
    "column7","column8")))

y
## column1 column2 column3 column4 column5 column6 column7 column8
## row1 1 4 7 10 13 16 19 22
## row2 2 5 8 11 14 17 20 23
```

```
## row3 3 6 9 12 15 18 21 24
y<-matrix(1:24, nrow = 3, ncol = 8,
  dimnames = list(row_names = c("row1","row2","row3"),
    column_names = c("column1","column2","column3",
      "column4","column5","column6",
      "column7","column8")))
```

```
y
## column_names
## row_names column1 column2 column3 column4 column5 column6
column7 column8
## row1 1 4 7 10 13 16 19 22
## row2 2 5 8 11 14 17 20 23
## row3 3 6 9 12 15 18 21 24
```

In the first example above, we set row names to each row and column names to each column. In the second example, we set row names and column names in addition to a name for all rows and a name for all columns.

To change the matrix row names and matrix column names, we can use the dimnames() function.

```
y<-matrix(1:24, nrow = 3, ncol = 8,

  dimnames = list(c("row1","row2","row3"),

    c("column1","column2","column3",
    "column4","column5","column6",
    "column7","column8")))
```

```
y
## column1 column2 column3 column4 column5 column6 column7
column8
```

```
## row1 1 4 7 10 13 16 19 22
## row2 2 5 8 11 14 17 20 23
## row3 3 6 9 12 15 18 21 24
dimnames(y)<-list(rows = c("r1","r2","r3"),

  columns = c("c1","c2","c3",
   "c4","c5","c6",
   "c7","c8"))

y
## columns
## rows c1 c2 c3 c4 c5 c6 c7 c8
## r1 1 4 7 10 13 16 19 22
## r2 2 5 8 11 14 17 20 23
## r3 3 6 9 12 15 18 21 24
```

To set the names of columns or rows separately, use the colnames() or rownames() function.

```
y<-matrix(1:24, nrow = 3, ncol = 8,

 dimnames = list(c("row1","row2","row3"),

   c("column1","column2","column3",
   "column4","column5","column6",
   "column7","column8")))

y
## column1 column2 column3 column4 column5 column6 column7
column8
## row1 1 4 7 10 13 16 19 22
## row2 2 5 8 11 14 17 20 23
```

```
## row3 3 6 9 12 15 18 21 24
colnames(y)<-c("col1","col2","col3",
  "col4","col5","col6",
  "col7","col8")

y
##  col1 col2 col3 col4 col5 col6 col7 col8
## row1 1 4 7 10 13 16 19 22
## row2 2 5 8 11 14 17 20 23
## row3 3 6 9 12 15 18 21 24
```

In the above example, we changed the column names only.
In the below example, we will change the row names only.

```
y<-matrix(1:24, nrow = 3, ncol = 8,

  dimnames = list(c("row1","row2","row3"),

     c("column1","column2","column3",
     "column4","column5","column6",
     "column7","column8")))

y
##   column1 column2 column3 column4 column5 column6 column7
column8
## row1  1  4  7  10  13  16  19  22
## row2  2  5  8  11  14  17  20  23
## row3  3  6  9  12  15  18  21  24
rownames(y)<-c("r1","r2","r3")

y
```

## column1 column2 column3 column4 column5 column6 column7 column8

## r1 1 4 7 10 13 16 19 22

## r2 2 5 8 11 14 17 20 23

## r3 3 6 9 12 15 18 21 24

# 5.2 CBIND() AND RBIND() FUNCTIONS

## 5.2.1 Combining Vectors

cbind() functions can be used to create matrices by combining different vectors by columns.

```
x<–1:10

x
## [1] 1 2 3 4 5 6 7 8 9 10
y<–11:20

y
## [1] 11 12 13 14 15 16 17 18 19 20
l<–21:30

l
## [1] 21 22 23 24 25 26 27 28 29 30
z<-cbind(x,y)

z
## x y
## [1,] 1 11
## [2,] 2 12
## [3,] 3 13
## [4,] 4 14
## [5,] 5 15
## [6,] 6 16
## [7,] 7 17
## [8,] 8 18
## [9,] 9 19
## [10,] 10 20
```

```
z<-cbind(x,y,l)

z
## x y l
## [1,] 1 11 21
## [2,] 2 12 22
## [3,] 3 13 23
## [4,] 4 14 24
## [5,] 5 15 25
## [6,] 6 16 26
## [7,] 7 17 27
## [8,] 8 18 28
## [9,] 9 19 29
## [10,] 10 20 30
z<-cbind(l,x,y)

z
## l x y
## [1,] 21 1 11
## [2,] 22 2 12
## [3,] 23 3 13
## [4,] 24 4 14
## [5,] 25 5 15
## [6,] 26 6 16
## [7,] 27 7 17
## [8,] 28 8 18
## [9,] 29 9 19
## [10,] 30 10 20
```

The columns are combined in the order they were supplied to cbind() function.
rbind() functions can be used to create matrices by combining different vectors by rows.

```
x<-1:10

x
## [1] 1 2 3 4 5 6 7 8 9 10
```

```
y<–11:20

y
## [1] 11 12 13 14 15 16 17 18 19 20
l<–21:30

l
## [1] 21 22 23 24 25 26 27 28 29 30
z<-rbind(x,y)

z
## [,1] [,2] [,3] [,4] [,5] [,6] [,7] [,8] [,9] [,10]
## x 1 2 3 4 5 6 7 8 9 10
## y 11 12 13 14 15 16 17 18 19 20
z<-rbind(x,y,l)

z
## [,1] [,2] [,3] [,4] [,5] [,6] [,7] [,8] [,9] [,10]
## x 1 2 3 4 5 6 7 8 9 10
## y 11 12 13 14 15 16 17 18 19 20
## l 21 22 23 24 25 26 27 28 29 30
z<-rbind(l,x,y)

z
## [,1] [,2] [,3] [,4] [,5] [,6] [,7] [,8] [,9] [,10]
## l 21 22 23 24 25 26 27 28 29 30
## x 1 2 3 4 5 6 7 8 9 10
## y 11 12 13 14 15 16 17 18 19 20
```

The rows are combined in the order they were supplied to rbind() function. In the above examples, the vectors are of equal length. If one of the vectors combined, recycling will occur with warning if longer vectors are not multiple of shorter vectors.

```
x<–1:10

x
## [1] 1 2 3 4 5 6 7 8 9 10
y<–11:20
```

```
y
## [1] 11 12 13 14 15 16 17 18 19 20
z<-rbind(1,x,y)

z
## [,1] [,2] [,3] [,4] [,5] [,6] [,7] [,8] [,9] [,10]
##  1 1 1 1 1 1 1 1 1  1
## x 1 2 3 4 5 6 7 8 9 10
## y 11 12 13 14 15 16 17 18 19 20
z<-rbind(1:3,x,y)
## Warning in rbind(1:3, x, y): number of columns of result is not a multiple
of
## vector length (arg 1)
z
## [,1] [,2] [,3] [,4] [,5] [,6] [,7] [,8] [,9] [,10]
##  1 2 3 1 2 3 1 2 3  1
## x 1 2 3 4 5 6 7 8 9 10
## y 11 12 13 14 15 16 17 18 19 20
z<-rbind(1:5,x,y)

z
## [,1] [,2] [,3] [,4] [,5] [,6] [,7] [,8] [,9] [,10]
##  1 2 3 4 5 1 2 3 4  5
## x 1 2 3 4 5 6 7 8 9 10
## y 11 12 13 14 15 16 17 18 19 20
z<-cbind(1:5,x,y)

z
##   x y
## [1,] 1 1 11
## [2,] 2 2 12
## [3,] 3 3 13
## [4,] 4 4 14
## [5,] 5 5 15
## [6,] 1 6 16
## [7,] 2 7 17
## [8,] 3 8 18
## [9,] 4 9 19
```

```
## [10,] 5 10 20
```

The warning is only produced when a vector of length 3 is row binded with vectors of length 10.

### 5.2.2 Combining Matrices

rbind() and cbind() can also combine matrices by rows or columns.

USPersonalExpenditure is a built-in numerical matrix in R. This data set consists of United States personal expenditures (in billions of dollars) in the categories; food and tobacco, household operation, medical, and health, personal care, and private education for the years 1940, 1945, 1950, 1955 and 1960. It has 5 rows and 5 columns.

```
USPersonalExpenditure
##      1940 1945 1950 1955 1960
## Food and Tobacco 22.200 44.500 59.60 73.2 86.80
## Household Operation 10.500 15.500 29.00 36.5 46.20
## Medical and Health 3.530 5.760 9.71 14.0 21.10
## Personal Care  1.040 1.980 2.45 3.4 5.40
## Private Education 0.341 0.974 1.80 2.6 3.64
z<-cbind(USPersonalExpenditure,USPersonalExpenditure)

z
##      1940 1945 1950 1955 1960 1940 1945 1950 1955
## Food and Tobacco 22.200 44.500 59.60 73.2 86.80 22.200 44.500 59.60
73.2
## Household Operation 10.500 15.500 29.00 36.5 46.20 10.500 15.500
29.00 36.5
## Medical and Health 3.530 5.760 9.71 14.0 21.10 3.530 5.760 9.71 14.0
## Personal Care  1.040 1.980 2.45 3.4 5.40 1.040 1.980 2.45 3.4
## Private Education 0.341 0.974 1.80 2.6 3.64 0.341 0.974 1.80 2.6
##      1960
## Food and Tobacco 86.80
## Household Operation 46.20
## Medical and Health 21.10
## Personal Care  5.40
## Private Education 3.64
z<-rbind(USPersonalExpenditure,USPersonalExpenditure)
```

```
z
##     1940 1945 1950 1955 1960
## Food and Tobacco 22.200 44.500 59.60 73.2 86.80
## Household Operation 10.500 15.500 29.00 36.5 46.20
## Medical and Health 3.530 5.760 9.71 14.0 21.10
## Personal Care  1.040 1.980 2.45 3.4 5.40
## Private Education 0.341 0.974 1.80 2.6 3.64
## Food and Tobacco 22.200 44.500 59.60 73.2 86.80
## Household Operation 10.500 15.500 29.00 36.5 46.20
## Medical and Health 3.530 5.760 9.71 14.0 21.10
## Personal Care  1.040 1.980 2.45 3.4 5.40
## Private Education 0.341 0.974 1.80 2.6 3.64
```

If you have different data types supplied to rbind() or cbind(), coercion will occur.
Letters is a built-in character vector in R that contains the 26 small letters of the English language.
Letters is a built-in character vector in R that contains the 26 capital letters of the English language.
month.abb is a built-in character vector in R that contains the three-letter abbreviations for the English month names.
month.name is a built-in character vector in R that contains the English names for the months of the year.

```
letters
## [1] “a” “b” “c” “d” “e” “f” “g” “h” “i” “j” “k” “l” “m” “n” “o” “p” “q”
“r” “s”
## [20] “t” “u” “v” “w” “x” “y” “z”
LETTERS
## [1] “A” “B” “C” “D” “E” “F” “G” “H” “I” “J” “K” “L” “M” “N” “O”
“P” “Q” “R” “S”
## [20] “T” “U” “V” “W” “X” “Y” “Z”
month.name
## [1] “January” “February” “March”  “April”  “May”  “June”
## [7] “July”  “August” “September” “October” “November” “December”
month.abb
## [1] “Jan” “Feb” “Mar” “Apr” “May” “Jun” “Jul” “Aug” “Sep” “Oct”
“Nov” “Dec”
```

```
z<-cbind(letters, 1:26)

z
## letters
## [1,] "a" "1"
## [2,] "b" "2"
## [3,] "c" "3"
## [4,] "d" "4"
## [5,] "e" "5"
## [6,] "f" "6"
## [7,] "g" "7"
## [8,] "h" "8"
## [9,] "i" "9"
## [10,] "j" "10"
## [11,] "k" "11"
## [12,] "l" "12"
## [13,] "m" "13"
## [14,] "n" "14"
## [15,] "o" "15"
## [16,] "p" "16"
## [17,] "q" "17"
## [18,] "r" "18"
## [19,] "s" "19"
## [20,] "t" "20"
## [21,] "u" "21"
## [22,] "v" "22"
## [23,] "w" "23"
## [24,] "x" "24"
## [25,] "y" "25"
## [26,] "z" "26"
z<-rbind(letters, 1:26)

z
## [,1] [,2] [,3] [,4] [,5] [,6] [,7] [,8] [,9] [,10] [,11] [,12] [,13]
## letters "a" "b" "c" "d" "e" "f" "g" "h" "i" "j" "k" "l" "m"
## "1" "2" "3" "4" "5" "6" "7" "8" "9" "10" "11" "12" "13"
## [,14] [,15] [,16] [,17] [,18] [,19] [,20] [,21] [,22] [,23] [,24] [,25]
## letters "n" "o" "p" "q" "r" "s" "t" "u" "v" "w" "x" "y"
## "14" "15" "16" "17" "18" "19" "20" "21" "22" "23" "24" "25"
```

```
## [,26]
## letters "z"
## "26"
z<-rbind(month.name, 1:12)

z
## [,1] [,2] [,3] [,4] [,5] [,6] [,7] [,8]
## month.name "January" "February" "March" "April" "May" "June"
"July" "August"
## "1" "2" "3" "4" "5" "6" "7" "8"
## [,9] [,10] [,11] [,12]
## month.name "September" "October" "November" "December"
## "9" "10" "11" "12"
z<-rbind(month.abb, 1:12)

z
## [,1] [,2] [,3] [,4] [,5] [,6] [,7] [,8] [,9] [,10] [,11]
## month.abb "Jan" "Feb" "Mar" "Apr" "May" "Jun" "Jul" "Aug" "Sep"
"Oct" "Nov"
## "1" "2" "3" "4" "5" "6" "7" "8" "9" "10" "11"
## [,12]
## month.abb "Dec"
## "12"
```

In all cases, the numeric data is converted to character data as evident from the double quotes around each number after row or column binding.

### 5.3 dim() FUNCTION

dim() function is used to get the dimensions of a matrix (and dataframes). The first number is the number of rows and the second number is the number of columns.

```
y<-matrix(1:24, nrow = 3, ncol = 8)

y
## [,1] [,2] [,3] [,4] [,5] [,6] [,7] [,8]
## [1,] 1 4 7 10 13 16 19 22
## [2,] 2 5 8 11 14 17 20 23
## [3,] 3 6 9 12 15 18 21 24
```

**dim**(y)
## [1] 3 8

3 is the number of rows and 8 is the number of columns in the y matrix. dim() also can create a matrix from a vector by setting these two numbers (the number of rows, the number of columns).

y<–1:24

y
## [1] 1 2 3 4 5 6 7 8 9 10 11 12 13 14 15 16 17 18 19 20 21 22 23 24
**dim**(y)<-**c**(3,8)

y
## [,1] [,2] [,3] [,4] [,5] [,6] [,7] [,8]
## [1,] 1 4 7 10 13 16 19 22
## [2,] 2 5 8 11 14 17 20 23
## [3,] 3 6 9 12 15 18 21 24
**dim**(y)<-**c**(4,6)

y
## [,1] [,2] [,3] [,4] [,5] [,6]
## [1,] 1 5 9 13 17 21
## [2,] 2 6 10 14 18 22
## [3,] 3 7 11 15 19 23
## [4,] 4 8 12 16 20 24
**dim**(y)<-**c**(6,4)

y
## [,1] [,2] [,3] [,4]
## [1,] 1 7 13 19
## [2,] 2 8 14 20
## [3,] 3 9 15 21
## [4,] 4 10 16 22
## [5,] 5 11 17 23
## [6,] 6 12 18 24
month.name
## [1] "January" "February" "March" "April" "May" "June"

## [7] "July" "August" "September" "October" "November" "December"
**dim**(month.name)<-**c**(3,4)

month.name
## [,1] [,2] [,3] [,4]
## [1,] "January" "April" "July" "October"
## [2,] "February" "May" "August" "November"
## [3,] "March" "June" "September" "December"

Note that matrix created are filled column-wise.
As noted above, dim() can also change the dimensions of already created matrices.

# 5.4 DATA.FRAME() FUNCTION

Dataframes are used to store tabular data in R like matrices with columns and rows.
Unlike matrices, data frames can store different classes of objects in each column. Matrices must have every element or column of the same class (e.g., all characters or all numeric).
Data frames are special type of list where every element of the list or dataframe is a column. All columns of the dataframes have equal length and this length is the number of rows.
Data frames are usually created by reading (in R) a dataset using the read.table(), read.csv() or read_xlsx() functions for example.
However, data frames can also be created explicitly with the data.frame() function.

df<-**data.frame**(ID = 1:5, name = **c**("Ann","John","Mary","Tom","Vector"),

blood_pressure = **c**(120,150,110,120,135),

Age = **c**(20,50,35,30,40))

df
## ID name blood_pressure Age
## 1 1 Ann 120 20
## 2 2 John 150 50

```
## 3 3 Mary    110 35
## 4 4 Tom    120 30
## 5 5 Vector    135 40
```

In the above example, we created an imaginary dataframe about some patients from scratch with 4 columns and 5 rows.

The ID is a numeric column containing ID for each patient. The name column is a character column containing the patient names. The blood_pressure is a numeric column that contains the blood pressure readings. The age is a numeric column that contains the age of each patient.

As a rule of thumb for tabular data, rows should contain the observations (or experimental units) and columns should contain the variables measured for these observations or experimental units (age, blood pressure, height, weight,…)

For dataframes, the names() function is used to get or change the column names (as lists).

```
df<-data.frame(ID = 1:5, name = c("Ann","John","Mary","Tom","Vector"),

  blood_pressure = c(120,150,110,120,135),

  Age = c(20,50,35,30,40))


df
## ID name blood_pressure Age
## 1 1 Ann    120 20
## 2 2 John    150 50
## 3 3 Mary    110 35
## 4 4 Tom    120 30
## 5 5 Vector    135 40
names(df)
## [1] "ID"    "name"    "blood_pressure" "Age"
names(df)<-c("serial","name","BP","Age (years)")
```

```
df
## serial name BP Age (years)
## 1  1 Ann 120   20
## 2  2 John 150   50
## 3  3 Mary 110   35
## 4  4 Tom 120   30
## 5  5 Vector 135   40
```

For dataframes, the row.names() function is used to get or change the row names.

By default, the row names of any dataframe in R is a character vector of the number of rows of this dataframe.

```
df<-data.frame(ID = 1:5, name = c("Ann","John","Mary","Tom","Vector"),

   blood_pressure = c(120,150,110,120,135),

   Age = c(20,50,35,30,40))

df
## ID name blood_pressure Age
## 1 1 Ann    120 20
## 2 2 John    150 50
## 3 3 Mary    110 35
## 4 4 Tom    120 30
## 5 5 Vector    135 40
row.names(df)
## [1] "1" "2" "3" "4" "5"
row.names(df)<-c("r1","r2","r3","r4","r5")

df
```

## ID name blood_pressure Age

## r1 1 Ann 120 20

## r2 2 John 150 50

## r3 3 Mary 110 35

## r4 4 Tom 120 30

## r5 5 Vector 135 40

## 5.5 EXAMINING THE STRUCTURE OF BUILT IN R DATAFRAMES

R programming language has many functions to explore the structure of large datasets.

We will examine these functions on some built-in dataframes in R.

To load any built-in dataframe in R, use the data() function.

### 5.5.1 Iris Data

This iris data set gives the measurements in centimeters of the variables sepal length and width and petal length and width, respectively, for 50 flowers from each of 3 species of iris. The species are *Iris setosa*, versicolor, and virginica.

*Reference:* The data were collected by Anderson (1935).

str() function gives the general structure of this dataframe. The str() function can also be used for any R object to examine its structure.

**data**("iris")

**str**(iris)

## 'data.frame': 150 obs. of 5 variables:

## $ Sepal.Length: num 5.1 4.9 4.7 4.6 5 5.4 4.6 5 4.4 4.9 ...

## $ Sepal.Width : num 3.5 3 3.2 3.1 3.6 3.9 3.4 3.4 2.9 3.1 ...

## $ Petal.Length: num 1.4 1.4 1.3 1.5 1.4 1.7 1.4 1.5 1.4 1.5 ...

## $ Petal.Width : num 0.2 0.2 0.2 0.2 0.2 0.4 0.3 0.2 0.2 0.1 ...

## $ Species : Factor w/ 3 levels "setosa","versicolor",..: 1 1 1 1 1 1 1 1 1 1 ...

str() function gives the following information:

1. iris is a dataframe with 150 obs.(observations or rows) and 5 variables (columns).
2. Sepal.length is the first column and it is a numeric vector. It then gives the first values of this column (5.1, 4.9, 4.7, 4.6, 5, 5.4, 4.6, 5, 4.4, 4.9, …).
3. Sepal.Width is the second column and it is a numeric vector. It then gives the first values of this column.
4. Petal.Length is the third column and it is a numeric vector. It then gives the first values of this column.
5. Petal.Width is the fourth column and it is a numeric vector. It then gives the first values of this column.
6. Species is the fifth column and it is a factor or categorical vector (see the next chapter). It then gives the first values of this column.

summary() function gives a summary for each column of this dataframe.

```
summary(iris)
## Sepal.Length Sepal.Width  Petal.Length Petal.Width
## Min. :4.300 Min. :2.000 Min. :1.000 Min. :0.100
## 1st Qu.:5.100 1st Qu.:2.800 1st Qu.:1.600 1st Qu.:0.300
## Median :5.800 Median :3.000 Median :4.350 Median :1.300
## Mean :5.843 Mean :3.057 Mean :3.758 Mean :1.199
## 3rd Qu.:6.400 3rd Qu.:3.300 3rd Qu.:5.100 3rd Qu.:1.800
## Max. :7.900 Max. :4.400 Max. :6.900 Max. :2.500
##  Species
## setosa :50
## versicolor:50
## virginica :50
##
##
##
```

summary() function gives us the following information:

1. Minimum, 1st quantile (25%), median (2nd quantile or 50%), mean, 3rd quantile (75%), and maximum values for numeric columns.
2. For factor columns, it gives the frequency of levels or categories of that column. In our case, there are 50 rows containing the

species setosa, 50 rows containing the species versicolor, and 50 rows containing the species virginica.

head() function gives the first 6 rows of our dataframe by default. You can increase the number of showed rows by passing a second argument greater than 6.

```
head(iris)
## Sepal.Length Sepal.Width Petal.Length Petal.Width Species
## 1  5.1  3.5  1.4  0.2 setosa
## 2  4.9  3.0  1.4  0.2 setosa
## 3  4.7  3.2  1.3  0.2 setosa
## 4  4.6  3.1  1.5  0.2 setosa
## 5  5.0  3.6  1.4  0.2 setosa
## 6  5.4  3.9  1.7  0.4 setosa
# show 1st 10 rows

head(iris,10)
## Sepal.Length Sepal.Width Petal.Length Petal.Width Species
## 1  5.1  3.5  1.4  0.2 setosa
## 2  4.9  3.0  1.4  0.2 setosa
## 3  4.7  3.2  1.3  0.2 setosa
## 4  4.6  3.1  1.5  0.2 setosa
## 5  5.0  3.6  1.4  0.2 setosa
## 6  5.4  3.9  1.7  0.4 setosa
## 7  4.6  3.4  1.4  0.3 setosa
## 8  5.0  3.4  1.5  0.2 setosa
## 9  4.4  2.9  1.4  0.2 setosa
## 10  4.9  3.1  1.5  0.1 setosa
# show 1st 15 rows

head(iris,15)
## Sepal.Length Sepal.Width Petal.Length Petal.Width Species
## 1  5.1  3.5  1.4  0.2 setosa
```

```
## 2  4.9  3.0  1.4  0.2 setosa
## 3  4.7  3.2  1.3  0.2 setosa
## 4  4.6  3.1  1.5  0.2 setosa
## 5  5.0  3.6  1.4  0.2 setosa
## 6  5.4  3.9  1.7  0.4 setosa
## 7  4.6  3.4  1.4  0.3 setosa
## 8  5.0  3.4  1.5  0.2 setosa
## 9  4.4  2.9  1.4  0.2 setosa
## 10  4.9  3.1  1.5  0.1 setosa
## 11  5.4  3.7  1.5  0.2 setosa
## 12  4.8  3.4  1.6  0.2 setosa
## 13  4.8  3.0  1.4  0.1 setosa
## 14  4.3  3.0  1.1  0.1 setosa
## 15  5.8  4.0  1.2  0.2 setosa
```

In the above examples, we showed the first 6, 10, and 15 rows respectively. tail() function gives the last 6 rows of our dataframe by default. You can increase the number of showed rows by passing a second argument greater than 6.

```
tail(iris)
##  Sepal.Length Sepal.Width Petal.Length Petal.Width Species
## 145  6.7  3.3  5.7  2.5 virginica
## 146  6.7  3.0  5.2  2.3 virginica
## 147  6.3  2.5  5.0  1.9 virginica
## 148  6.5  3.0  5.2  2.0 virginica
## 149  6.2  3.4  5.4  2.3 virginica
## 150  5.9  3.0  5.1  1.8 virginica
# show last 10 rows

tail(iris,10)
##  Sepal.Length Sepal.Width Petal.Length Petal.Width Species
```

```
## 141 6.7 3.1 5.6 2.4 virginica
## 142 6.9 3.1 5.1 2.3 virginica
## 143 5.8 2.7 5.1 1.9 virginica
## 144 6.8 3.2 5.9 2.3 virginica
## 145 6.7 3.3 5.7 2.5 virginica
## 146 6.7 3.0 5.2 2.3 virginica
## 147 6.3 2.5 5.0 1.9 virginica
## 148 6.5 3.0 5.2 2.0 virginica
## 149 6.2 3.4 5.4 2.3 virginica
## 150 5.9 3.0 5.1 1.8 virginica
# show last 15 rows

tail(iris,15)
## Sepal.Length Sepal.Width Petal.Length Petal.Width Species
## 136 7.7 3.0 6.1 2.3 virginica
## 137 6.3 3.4 5.6 2.4 virginica
## 138 6.4 3.1 5.5 1.8 virginica
## 139 6.0 3.0 4.8 1.8 virginica
## 140 6.9 3.1 5.4 2.1 virginica
## 141 6.7 3.1 5.6 2.4 virginica
## 142 6.9 3.1 5.1 2.3 virginica
## 143 5.8 2.7 5.1 1.9 virginica
## 144 6.8 3.2 5.9 2.3 virginica
## 145 6.7 3.3 5.7 2.5 virginica
## 146 6.7 3.0 5.2 2.3 virginica
## 147 6.3 2.5 5.0 1.9 virginica
## 148 6.5 3.0 5.2 2.0 virginica
## 149 6.2 3.4 5.4 2.3 virginica
## 150 5.9 3.0 5.1 1.8 virginica
```

### 5.5.2 Air Quality Data

It is the daily air quality measurements in New York, May to September 1973.

It contains the daily readings of the following air quality values:

1. **Ozone:** Mean ozone in parts per billion from 1300 to 1500 hours at Roosevelt Island
2. **Solar.R:** Solar radiation in Langleys in the frequency band 4000–7700 Angstroms from 0800 to 1200 hours at Central Park
3. **Wind:** Average wind speed in miles per hour at 0700 and 1000 hours at LaGuardia Airport (LGA).
4. **Temp:** Maximum daily temperature in degrees Fahrenheit at La Guardia Airport.

*Reference:* Chambers, Cleveland, Kleiner, and Tukey (1983).
str() function gives the general structure of this dataframe.
**data**("airquality")

**str**(airquality)

## 'data.frame': 153 obs. of 6 variables:

## $ Ozone : int 41 36 12 18 NA 28 23 19 8 NA ...

## $ Solar.R: int 190 118 149 313 NA NA 299 99 19 194 ...

## $ Wind : num 7.4 8 12.6 11.5 14.3 14.9 8.6 13.8 20.1 8.6 ...

## $ Temp : int 67 72 74 62 56 66 65 59 61 69 ...

## $ Month : int 5 5 5 5 5 5 5 5 5 5 ...

## $ Day : int 1 2 3 4 5 6 7 8 9 10 ...

str() function gives the following information:

1. airquality is a dataframe with 153 obs.(observations or rows) and 6 variables (columns).
2. Ozone is the first column and it is an integer vector. It then gives the first values of this column (41, 36, 12, 18, NA, 28, 23, 19, 8, NA ...). NA is a missing value.
3. Solar.R is the second column and it is an integer vector. It then gives the first values of this column.

4. Wind is the third column and it is a numeric vector. It then gives the first values of this column.
5. Temp is the fourth column and it is an integer vector. It then gives the first values of this column.
6. Month is the fifth column and it is an integer vector. It then gives the first values of this column.
7. Day is the sixth column and it is an integer vector. It then gives the first values of this column.

summary() function gives a summary for each column of this dataframe.

```
summary(airquality)
## Ozone Solar.R Wind Temp
## Min. : 1.00 Min. : 7.0 Min. : 1.700 Min. :56.00
## 1st Qu.: 18.00 1st Qu.:115.8 1st Qu.: 7.400 1st Qu.:72.00
## Median : 31.50 Median :205.0 Median : 9.700 Median :79.00
## Mean : 42.13 Mean :185.9 Mean : 9.958 Mean :77.88
## 3rd Qu.: 63.25 3rd Qu.:258.8 3rd Qu.:11.500 3rd Qu.:85.00
## Max. :168.00 Max. :334.0 Max. :20.700 Max. :97.00
## NA's :37 NA's :7
## Month Day
## Min. :5.000 Min. : 1.0
## 1st Qu.:6.000 1st Qu.: 8.0
## Median :7.000 Median :16.0
## Mean :6.993 Mean :15.8
## 3rd Qu.:8.000 3rd Qu.:23.0
## Max. :9.000 Max. :31.0
##
```

summary() function gives us the following information:

1. Minimum, 1st quantile (25%), median (2nd quantile or 50%), mean, 3rd quantile (75%), maximum, and NA (missing) values for numeric columns.

2. For factor columns, Month, and Day, summary() function treats them as a numeric column because they are integer columns. However, you

can obtain the frequency of levels or categories of these columns by using the factor() function. See the next chapter.

head() function gives the first 6 rows of our dataframe by default. You can increase the number of showed rows by passing a second argument greater than 6.

```
head(airquality)
## Ozone Solar.R Wind Temp Month Day
## 1 41  190 7.4 67  5 1
## 2 36  118 8.0 72  5 2
## 3 12  149 12.6 74  5 3
## 4 18  313 11.5 62  5 4
## 5 NA  NA 14.3 56  5 5
## 6 28  NA 14.9 66  5 6
# show 1st 10 rows

head(airquality,10)
## Ozone Solar.R Wind Temp Month Day
## 1  41  190 7.4 67  5 1
## 2  36  118 8.0 72  5 2
## 3  12  149 12.6 74  5 3
## 4  18  313 11.5 62  5 4
## 5  NA  NA 14.3 56  5 5
## 6  28  NA 14.9 66  5 6
## 7  23  299 8.6 65  5 7
## 8  19  99 13.8 59  5 8
## 9  8  19 20.1 61  5 9
## 10 NA  194 8.6 69  5 10
# show 1st 15 rows

head(airquality,15)
## Ozone Solar.R Wind Temp Month Day
```

```
## 1 41 190 7.4 67 5 1
## 2 36 118 8.0 72 5 2
## 3 12 149 12.6 74 5 3
## 4 18 313 11.5 62 5 4
## 5 NA NA 14.3 56 5 5
## 6 28 NA 14.9 66 5 6
## 7 23 299 8.6 65 5 7
## 8 19 99 13.8 59 5 8
## 9 8 19 20.1 61 5 9
## 10 NA 194 8.6 69 5 10
## 11 7 NA 6.9 74 5 11
## 12 16 256 9.7 69 5 12
## 13 11 290 9.2 66 5 13
## 14 14 274 10.9 68 5 14
## 15 18 65 13.2 58 5 15
```

In the above examples, we showed the first 6, 10, and 15 rows respectively tail() function gives the last 6 rows of our dataframe by default. You can increase the number of showed rows by passing a second argument greater than 6.

```
tail(airquality)
## Ozone Solar.R Wind Temp Month Day
## 148 14 20 16.6 63 9 25
## 149 30 193 6.9 70 9 26
## 150 NA 145 13.2 77 9 27
## 151 14 191 14.3 75 9 28
## 152 18 131 8.0 76 9 29
## 153 20 223 11.5 68 9 30
# show last 10 rows

tail(airquality,10)
```

```
## Ozone Solar.R Wind Temp Month Day
## 144 13 238 12.6 64 9 21
## 145 23 14 9.2 71 9 22
## 146 36 139 10.3 81 9 23
## 147 7 49 10.3 69 9 24
## 148 14 20 16.6 63 9 25
## 149 30 193 6.9 70 9 26
## 150 NA 145 13.2 77 9 27
## 151 14 191 14.3 75 9 28
## 152 18 131 8.0 76 9 29
## 153 20 223 11.5 68 9 30
# show last 15 rows

tail(airquality,15)
## Ozone Solar.R Wind Temp Month Day
## 139 46 237 6.9 78 9 16
## 140 18 224 13.8 67 9 17
## 141 13 27 10.3 76 9 18
## 142 24 238 10.3 68 9 19
## 143 16 201 8.0 82 9 20
## 144 13 238 12.6 64 9 21
## 145 23 14 9.2 71 9 22
## 146 36 139 10.3 81 9 23
## 147 7 49 10.3 69 9 24
## 148 14 20 16.6 63 9 25
## 149 30 193 6.9 70 9 26
## 150 NA 145 13.2 77 9 27
## 151 14 191 14.3 75 9 28
## 152 18 131 8.0 76 9 29
## 153 20 223 11.5 68 9 30
```

CHAPTER 6

# FACTORS AND MISSING VALUES

## CONTENTS

## 6.1 FACTOR() FUNCTION

Factors are used to represent categorical data and can be unordered or ordered.

Ordered factors like month names (January < February < March<.........< December) and weekdays (Monday < Tuesday < Wednesday<.........< Sunday).

Factors can be created with factor() from any vector type (numeric, integer, character, logical).

```
# numeric vector

x<–1:10

x
## [1] 1 2 3 4 5 6 7 8 9 10
x<-factor(x)

x
## [1] 1 2 3 4 5 6 7 8 9 10
## Levels: 1 2 3 4 5 6 7 8 9 10
# integer vector

x<–1L:10L

x
## [1] 1 2 3 4 5 6 7 8 9 10
x<-factor(x)

x
## [1] 1 2 3 4 5 6 7 8 9 10
## Levels: 1 2 3 4 5 6 7 8 9 10
# character vector

x<-month.abb

x
## [1] "Jan" "Feb" "Mar" "Apr" "May" "Jun" "Jul" "Aug" "Sep" "Oct"
"Nov" "Dec"
```

```
x<-factor(x)

x
## [1] Jan Feb Mar Apr May Jun Jul Aug Sep Oct Nov Dec
## Levels: Apr Aug Dec Feb Jan Jul Jun Mar May Nov Oct Sep
# logical vector

x<–1:10

x
## [1] 1 2 3 4 5 6 7 8 9 10
y<- x>5

y
## [1] FALSE FALSE FALSE FALSE FALSE TRUE TRUE TRUE TRUE
TRUE
y<-factor(y)

y
## [1] FALSE FALSE FALSE FALSE FALSE TRUE TRUE TRUE TRUE
TRUE
## Levels: FALSE TRUE
```

In the factor() function output, there are levels. Levels are the unique values for that vector. The first value in the levels is the reference value for which all other values are compared.

The order of levels is important for statistical tests or modeling to define the reference level to which all other levels are compared. It is also important in plotting where the reference (first) level will be plotted as the first bar in bar or column plots for categorical variables.

R, by default, sorts the levels alphabetically, or if they are numbers, they are sorted in increasing order as shown in the examples above.

In the example of month.abb, the result is not what we want. To change the order of levels, use the levels argument within the factor() function.

```
x<-month.abb

x
```

```
## [1] "Jan" "Feb" "Mar" "Apr" "May" "Jun" "Jul" "Aug" "Sep" "Oct"
"Nov" "Dec"
x<-factor(x)

x
## [1] Jan Feb Mar Apr May Jun Jul Aug Sep Oct Nov Dec
## Levels: Apr Aug Dec Feb Jan Jul Jun Mar May Nov Oct Sep
x<-factor(x, levels = c("Jan","Feb","Mar","Apr","May","Jun",
    "Jul","Aug","Sep","Oct","Nov","Dec"))

x
## [1] Jan Feb Mar Apr May Jun Jul Aug Sep Oct Nov Dec
## Levels: Jan Feb Mar Apr May Jun Jul Aug Sep Oct Nov Dec
```

Another important argument is the labels() which can be used to give labels to your levels. In our example of month.abb, we can add labels of numbers to indicate the month name. We can also add duplicated values to indicate quarters.

Using factors with self-describing labels is better than using integers. Having a variable that has levels "Male" and "Female" is better than a variable that has values 1 and 2.

```
x<-month.abb

x
## [1] "Jan" "Feb" "Mar" "Apr" "May" "Jun" "Jul" "Aug" "Sep" "Oct"
"Nov" "Dec"
x<-factor(x, levels = c("Jan","Feb","Mar","Apr","May","Jun",
    "Jul","Aug","Sep","Oct","Nov","Dec"))

x
## [1] Jan Feb Mar Apr May Jun Jul Aug Sep Oct Nov Dec
## Levels: Jan Feb Mar Apr May Jun Jul Aug Sep Oct Nov Dec
# add numeric labels

x<-factor(x, levels = c("Jan","Feb","Mar","Apr","May","Jun",
    "Jul","Aug","Sep","Oct","Nov","Dec"),

  labels = 1:12)
```

```
x
## [1] 1 2 3 4 5 6 7 8 9 10 11 12
## Levels: 1 2 3 4 5 6 7 8 9 10 11 12
## add quarter labels

x<-month.abb

x
## [1] “Jan” “Feb” “Mar” “Apr” “May” “Jun” “Jul” “Aug” “Sep” “Oct”
“Nov” “Dec”
x<-factor(x, levels = c(“Jan”,”Feb”,”Mar”,”Apr”,”May”,”Jun”,
    “Jul”,”Aug”,”Sep”,”Oct”,”Nov”,”Dec”),

  labels = c(“Q1”, “Q1”, “Q1”, “Q2”, “Q2”, “Q2”,
    “Q3”, “Q3”, “Q3”, “Q4”, “Q4”, “Q4”))

x
## [1] Q1 Q1 Q1 Q2 Q2 Q2 Q3 Q3 Q3 Q4 Q4 Q4
## Levels: Q1 Q2 Q3 Q4
```

Another argument is ordered argument. set ordered = TRUE if you want ordered factor. Ordered factors differ from factors only in their class, but methods and the model-fitting functions treat the two classes quite differently so be caution when using it.

```
x<-month.abb

x
## [1] “Jan” “Feb” “Mar” “Apr” “May” “Jun” “Jul” “Aug” “Sep” “Oct”
“Nov” “Dec”
x<-factor(x, levels = c(“Jan”,”Feb”,”Mar”,”Apr”,”May”,”Jun”,
    “Jul”,”Aug”,”Sep”,”Oct”,”Nov”,”Dec”),

  ordered = TRUE)

x
## [1] Jan Feb Mar Apr May Jun Jul Aug Sep Oct Nov Dec
## 12 Levels: Jan < Feb < Mar < Apr < May < Jun < Jul < Aug < Sep < ...
```

< Dec

## 6.2 TABLE() AND PROP.TABLE() FUNCTIONS

table() function uses the cross-classifying factors to build a contingency table of the counts at each combination of factor levels.

### 6.2.1 Table() Function for a Single Column

In the airquality data, the Month and Day columns are integer vectors, but the table() function can also be applied. To extract a certain column from a dataframe (or a list), use the dollar sign "$".

The first row of the output is the categories' name and the second row is the row (observation) count for that category.

```
data("airquality")


str(airquality)
## 'data.frame': 153 obs. of 6 variables:
## $ Ozone : int 41 36 12 18 NA 28 23 19 8 NA ...
## $ Solar.R: int 190 118 149 313 NA NA 299 99 19 194 ...
## $ Wind : num 7.4 8 12.6 11.5 14.3 14.9 8.6 13.8 20.1 8.6 ...
## $ Temp : int 67 72 74 62 56 66 65 59 61 69 ...
## $ Month : int 5 5 5 5 5 5 5 5 5 5 ...
## $ Day : int 1 2 3 4 5 6 7 8 9 10 ...
table(airquality$Month)
##
## 5 6 7 8 9
## 31 30 31 31 30
summary(factor(airquality$Month))
## 5 6 7 8 9
## 31 30 31 31 30
table(airquality$Day)
##
## 1 2 3 4 5 6 7 8 9 10 11 12 13 14 15 16 17 18 19 20 21 22 23 24 25 26
## 5 5 5 5 5 5 5 5 5 5 5 5 5 5 5 5 5 5 5 5 5 5 5 5 5 5
```

```
## 27 28 29 30 31
## 5 5 5 5 3
summary(factor(airquality$Day))
## 1 2 3 4 5 6 7 8 9 10 11 12 13 14 15 16 17 18 19 20 21 22 23 24 25 26
## 5 5 5 5 5 5 5 5 5 5 5 5 5 5 5 5 5 5 5 5 5 5 5 5 5 5
## 27 28 29 30 31
## 5 5 5 5 3
```

In the examples above, using the table() function for the Month column showed that there are 31 rows for month 5 (May), 30 rows for month 6 (June), 31 rows for month 7 (July), 31 rows for month 8 (August), and 30 rows for month 9 (September). The total is 153 rows which is the total number of rows of this dataframe. Summary(factor(airquality$Month)) gave the same result.

Using the table() function for the Day column showed that there are 5 rows for each day number from 1 to 30 (1,2,3,4,…,30) except for 31 where there are only 3 rows. The total is 153 rows which is the total number of rows of this dataframe. Summary(factor(airquality$Day)) gave the same result.

Although table() and summary(factor) functions gave the same results, when there are NA values, they are not returned by table() function except when the argument, exclude=NULL or argument, useNA = "ifany" is used.

```
table(airquality$Ozone)
##
## 1 4 6 7 8 9 10 11 12 13 14 16 18 19 20 21 22 23 24 27
## 1 1 1 3 1 3 1 3 2 4 4 4 4 1 4 4 1 6 2 1
## 28 29 30 31 32 34 35 36 37 39 40 41 44 45 46 47 48 49 50 52
## 3 1 2 1 3 1 2 2 2 2 1 1 3 2 1 1 1 1 1 1
## 59 61 63 64 65 66 71 73 76 77 78 79 80 82 84 85 89 91 96 97
## 2 1 1 2 1 1 1 2 1 1 2 1 1 1 1 2 1 1 1 2
## 108 110 115 118 122 135 168
## 1 1 1 1 1 1 1
summary(factor(airquality$Ozone))
```

```
## 1 4 6 7 8 9 10 11 12 13 14 16 18 19 20 21
## 1 1 1 3 1 3 1 3 2 4 4 4 4 1 4 4
## 22 23 24 27 28 29 30 31 32 34 35 36 37 39 40 41
## 1 6 2 1 3 1 2 1 3 1 2 2 2 2 1 1
## 44 45 46 47 48 49 50 52 59 61 63 64 65 66 71 73
## 3 2 1 1 1 1 1 1 2 1 1 2 1 1 1 2
## 76 77 78 79 80 82 84 85 89 91 96 97 108 110 115 118
## 1 1 2 1 1 1 1 2 1 1 1 2 1 1 1 1
## 122 135 168 NA's
## 1 1 1 37
table(airquality$Ozone, exclude = NULL)
##
## 1 4 6 7 8 9 10 11 12 13 14 16 18 19 20 21
## 1 1 1 3 1 3 1 3 2 4 4 4 4 1 4 4
## 22 23 24 27 28 29 30 31 32 34 35 36 37 39 40 41
## 1 6 2 1 3 1 2 1 3 1 2 2 2 2 1 1
## 44 45 46 47 48 49 50 52 59 61 63 64 65 66 71 73
## 3 2 1 1 1 1 1 1 2 1 1 2 1 1 1 2
## 76 77 78 79 80 82 84 85 89 91 96 97 108 110 115 118
## 1 1 2 1 1 1 1 2 1 1 1 2 1 1 1 1
## 122 135 168 <NA>
## 1 1 1 37
table(airquality$Ozone, useNA = "ifany")
##
## 1 4 6 7 8 9 10 11 12 13 14 16 18 19 20 21
## 1 1 1 3 1 3 1 3 2 4 4 4 4 1 4 4
## 22 23 24 27 28 29 30 31 32 34 35 36 37 39 40 41
## 1 6 2 1 3 1 2 1 3 1 2 2 2 2 1 1
## 44 45 46 47 48 49 50 52 59 61 63 64 65 66 71 73
## 3 2 1 1 1 1 1 1 2 1 1 2 1 1 1 2
## 76 77 78 79 80 82 84 85 89 91 96 97 108 110 115 118
```

```
## 1 1 2 1 1 1 1 2 1 1 1 2 1 1 1 1
## 122 135 168 <NA>
## 1 1 1 37
```

Using the table() function only has not revealed the 37 NA values present in the Ozone column. These 37 NA values were showed by using summary(factor), table(exclude = NULL), or table(useNA = "ifany") functions.

### 6.2.2 Table() Function for Two Columns

```
table(airquality$Month,airquality$Day)
##
## 1 2 3 4 5 6 7 8 9 10 11 12 13 14 15 16 17 18 19 20 21 22 23 24 25 26
27 28
## 5 1 1 1 1 1 1 1 1 1 1 1 1 1 1 1 1 1 1 1 1 1 1 1 1 1 1 1 1
## 6 1 1 1 1 1 1 1 1 1 1 1 1 1 1 1 1 1 1 1 1 1 1 1 1 1 1 1 1
## 7 1 1 1 1 1 1 1 1 1 1 1 1 1 1 1 1 1 1 1 1 1 1 1 1 1 1 1 1
## 8 1 1 1 1 1 1 1 1 1 1 1 1 1 1 1 1 1 1 1 1 1 1 1 1 1 1 1 1
## 9 1 1 1 1 1 1 1 1 1 1 1 1 1 1 1 1 1 1 1 1 1 1 1 1 1 1 1 1
##
## 29 30 31
## 5 1 1 1
## 6 1 1 0
## 7 1 1 1
## 8 1 1 1
## 9 1 1 0
table(airquality$Day,airquality$Month)
##
## 5 6 7 8 9
## 1 1 1 1 1 1
## 2 1 1 1 1 1
## 3 1 1 1 1 1
```

```
## 4 1 1 1 1 1
## 5 1 1 1 1 1
## 6 1 1 1 1 1
## 7 1 1 1 1 1
## 8 1 1 1 1 1
## 9 1 1 1 1 1
## 10 1 1 1 1 1
## 11 1 1 1 1 1
## 12 1 1 1 1 1
## 13 1 1 1 1 1
## 14 1 1 1 1 1
## 15 1 1 1 1 1
## 16 1 1 1 1 1
## 17 1 1 1 1 1
## 18 1 1 1 1 1
## 19 1 1 1 1 1
## 20 1 1 1 1 1
## 21 1 1 1 1 1
## 22 1 1 1 1 1
## 23 1 1 1 1 1
## 24 1 1 1 1 1
## 25 1 1 1 1 1
## 26 1 1 1 1 1
## 27 1 1 1 1 1
## 28 1 1 1 1 1
## 29 1 1 1 1 1
## 30 1 1 1 1 1
## 31 1 0 1 1 0
```

Using the table() function with two columns, has created a crosstab with month as rows and days as columns for the first function, and day as rows and month as columns for the second function.

For example, there is one observation for day 1 and month 5 (May).

If you want to see NA values, you have to use the arguments, exclude = NULL) or useNA = "ifany", as before. Note that there is no NA values in these two columns.

Another example of using table() function for large datasets.

The regicor dataframe, from the package compareGroups, contains data from 3 different cross-sectional surveys of individuals representative of the population from a north-west Spanish province (Girona), REGICOR study. Visit [www.regicor.org].

R packages are a collection of R functions and sample data. These R functions makes your life easier by performing complex tasks that requires a lot of code lines.

By default, R installs a set of packages during installation. More packages are added later, when they are needed for some specific purpose. When we start the R console, only the default packages are available by default (all our previous functions come from these default packages). Other packages which are already installed have to be loaded explicitly by using library() function.

We will first install package compareGroups using install.packages() function, then activate the package in our working session using the library() function. Then load the regicor data in our working session using the data() function. As I have already installed that package in R so I put # before that line to not activate that code line again.

```
#install.packages("compareGroups")

library(compareGroups)
## Warning: package 'compareGroups' was built under R version 3.6.3
## Loading required package: SNPassoc
## Warning: package 'SNPassoc' was built under R version 3.6.3
## Loading required package: haplo.stats
## Warning: package 'haplo.stats' was built under R version 3.6.3
## Loading required package: survival
## Warning: package 'survival' was built under R version 3.6.3
## Loading required package: mvtnorm
```

```
## Loading required package: parallel
## Registered S3 method overwritten by 'SNPassoc':
## method   from
## summary.haplo.glm haplo.stats
data("regicor")

# data structure

str(regicor)
## 'data.frame': 2294 obs. of 25 variables:
## $ id  : num 2.26e+03 1.88e+03 3.00e+09 3.00e+09 3.00e+09 ...
## ..- attr(*, "label")= Named chr "Individual id"
## .. ..- attr(*, "names")= chr "id"
## $ year : Factor w/ 3 levels "1995","2000",..: 3 3 2 2 2 2 2 1 3 1 ...
## ..- attr(*, "label")= Named chr "Recruitment year"
## .. ..- attr(*, "names")= chr "year"
## $ age  : int 70 56 37 69 70 40 66 53 43 70 ...
## ..- attr(*, "label")= Named chr "Age"
## .. ..- attr(*, "names")= chr "age"
## $ sex  : Factor w/ 2 levels "Male","Female": 2 2 1 2 2 2 1 2 2 1 ...
## ..- attr(*, "label")= chr "Sex"
## $ smoker : Factor w/ 3 levels "Never smoker",..: 1 1 2 1 NA 2 1 1 3 3 ...
## ..- attr(*, "label")= Named chr "Smoking status"
## .. ..- attr(*, "names")= chr "smoker"
## $ sbp  : int 138 139 132 168 NA 108 120 132 95 142 ...
## ..- attr(*, "label")= Named chr "Systolic blood pressure"
## .. ..- attr(*, "names")= chr "sbp"
## $ dbp  : int 75 89 82 97 NA 70 72 78 65 78 ...
## ..- attr(*, "label")= Named chr "Diastolic blood pressure"
## .. ..- attr(*, "names")= chr "dbp"
## $ histhtn : Factor w/ 2 levels "Yes","No": 2 2 2 2 2 2 1 2 2 2 ...
```

```
## ..- attr(*, "label")= Named chr "History of hypertension"
## .. ..- attr(*, "names")= chr "histbp"
## $ txhtn : Factor w/ 2 levels "No","Yes": 1 1 1 1 1 1 2 1 1 1 ...
## ..- attr(*, "label")= chr "Hypertension treatment"
## $ chol : num 294 220 245 168 NA NA 298 254 194 188 ...
## ..- attr(*, "label")= Named chr "Total cholesterol"
## .. ..- attr(*, "names")= chr "chol"
## $ hdl  : num 57 50 59.8 53.2 NA ...
## ..- attr(*, "label")= Named chr "HDL cholesterol"
## .. ..- attr(*, "names")= chr "hdl"
## $ triglyc : num 93 160 89 116 NA 94 71 NA 68 137 ...
## ..- attr(*, "label")= Named chr "Triglycerides"
## .. ..- attr(*, "names")= chr "triglyc"
## $ ldl  : num 218.4 138 167.4 91.6 NA ...
## ..- attr(*, "label")= Named chr "LDL cholesterol"
## .. ..- attr(*, "names")= chr "ldl"
## $ histchol: Factor w/ 2 levels "Yes","No": 2 2 2 2 NA 2 1 2 2 2 ...
## ..- attr(*, "label")= chr "History of hyperchol."
## $ txchol : Factor w/ 2 levels "No","Yes": 1 1 1 1 NA 1 1 1 1 1 ...
## ..- attr(*, "label")= Named chr "Cholesterol treatment"
## .. ..- attr(*, "names")= chr "txchol"
## $ height : num 160 163 170 147 NA ...
## ..- attr(*, "label")= Named chr "Height (cm)"
## .. ..- attr(*, "names")= chr "height"
## $ weight : num 64 67 70 68 NA 43.5 79.2 45.8 53 62 ...
## ..- attr(*, "label")= Named chr "Weight (Kg)"
## .. ..- attr(*, "names")= chr "weight"
## $ bmi  : num 25 25.2 24.2 31.5 NA ...
## ..- attr(*, "label")= Named chr "Body mass index"
## .. ..- attr(*, "names")= chr "bmi"
## $ phyact : num 304 160 553 522 NA ...
```

```
## ..- attr(*, "label")= Named chr "Physical activity (Kcal/week)"
## .. ..- attr(*, "names")= chr "phyact"
## $ pcs  : num 54.5 58.2 43.4 54.3 NA ...
## ..- attr(*, "label")= Named chr "Physical component"
## .. ..- attr(*, "names")= chr "pcs"
## $ mcs  : num 58.9 48 62.6 57.9 NA ...
## ..- attr(*, "label")= chr "Mental component"
## $ cv  : Factor w/ 2 levels "No","Yes": 1 1 1 1 NA 1 1 1 1 1 ...
## ..- attr(*, "label")= chr "Cardiovascular event"
## $ tocv : num 1025 2757 1906 1055 NA ...
## ..- attr(*, "label")= chr "Days to cardiovascular event or end of follow-up"
## $ death : Factor w/ 2 levels "No","Yes": 2 1 1 1 NA 1 2 1 1 1 ...
## ..- attr(*, "label")= chr "Overall death"
## $ todeath : num 1299.2 39.3 858.4 1833.1 NA ...
## ..- attr(*, "label")= chr "Days to overall death or end of follow-up"
# first 6 rows

head(regicor)
##    id year age sex    smoker sbp dbp histhtn txhtn
## 6101  2265 2005 70 Female   Never smoker 138 75  No No
## 5762  1882 2005 56 Female   Never smoker 139 89  No No
## 2992 3000105616 2000 37 Male Current or former < 1y 132 82  No No
## 2611 3000103485 2000 69 Female   Never smoker 168 97  No No
## 2762 3000103963 2000 70 Female     <NA> NA NA  No No
## 1516 3000100883 2000 40 Female Current or former < 1y 108 70  No No
##  chol  hdl triglyc  ldl histchol txchol height weight  bmi
## 6101 294 57.00000  93 218.40000  No  No 160 64.0 25.00000
## 5762 220 50.00000  160 138.00000  No  No 163 67.0 25.21736
## 2992 245 59.80429  89 167.39571  No  No 170 70.0 24.22145
## 2611 168 53.17571  116 91.62429  No  No 147 68.0 31.46837
```

```
## 2762 NA NA NA NA <NA> <NA> NA NA NA
## 1516 NA 68.90000 94 NA No No 158 43.5 17.42509
## phyact pcs mcs cv tocv death todeath
## 6101 304.2000 54.455 58.918 No 1024.882 Yes 1299.16343
## 5762 160.3000 58.165 47.995 No 2756.849 No 39.32629
## 2992 552.7912 43.429 62.585 No 1905.969 No 858.42203
## 2611 522.0000 54.325 57.900 No 1055.380 No 1833.07619
## 2762 NA NA NA <NA> NA <NA> NA
## 1516 386.9505 57.315 47.869 No 3239.241 No 877.61155
# last 6 rows

tail(regicor)
## id year age sex smoker sbp dbp histhtn txhtn
## 7334 3672 2005 58 Male Former >= 1y 110 77 No No
## 62 1000001076 1995 42 Male Current or former < 1y 112 64 No No
## 3863 3000107763 2000 68 Male Never smoker 120 80 Yes No
## 3686 3000107330 2000 64 Male Never smoker 198 110 No No
## 4335 258 2005 49 Female Never smoker 155 100 No No
## 6786 3032 2005 42 Male Never smoker 107 65 No No


## chol hdl triglyc ldl histchol txchol height weight bmi
## 7334 235 50.00000 93 166.4000 Yes No 165 62.0 22.77319
## 62 215 42.30000 118 149.1000 Yes No 178 68.0 21.46194
## 3863 191 50.89000 91 121.9100 Yes No 170 80.0 27.68166
## 3686 237 46.43286 122 166.1671 No No 170 82.5 28.54671
## 4335 186 37.00000 155 118.0000 No No 158 67.0 26.83865
## 6786 243 71.00000 89 154.2000 No No 165 66.0 24.24242
## phyact pcs mcs cv tocv death todeath
## 7334 346.7000 59.49900 39.28400 No 3617.2435 No 441.3844
## 62 NA 58.01476 35.03757 No 1451.6946 No 3487.0368
```

```
## 3863 389.6703 51.90300 55.14400 No 1003.7665 No 836.4889
## 3686 157.1209 50.36100 51.26000 No 1220.7234 Yes 1010.7434
## 4335 1093.5000 57.63100 42.67800 No 331.7726 No 3133.9241
## 6786 353.7000 50.46100 57.43000 No 698.8185 No 1899.4416
# data summary

summary(regicor)
## id   year   age   sex
## Min. :2.000e+00 1995: 431 Min. :35.00 Male :1101
## 1st Qu.:2.128e+03 2000: 786 1st Qu.:46.00 Female:1193
## Median :1.000e+09 2005:1077 Median :55.00
## Mean :1.216e+09    Mean :54.74
## 3rd Qu.:3.000e+09    3rd Qu.:64.00
## Max. :3.000e+09    Max. :74.00
##
##     smoker   sbp   dbp   histhtn
## Never smoker   :1201 Min. : 80.0 Min. : 40.00 Yes : 723
## Current or former < 1y: 593 1st Qu.:116.0 1st Qu.: 72.00 No :1563
## Former >= 1y   : 439 Median :129.0 Median : 80.00 NA's: 8
## NA's    : 61 Mean :131.2 Mean : 79.66
##       3rd Qu.:144.0 3rd Qu.: 86.00
##       Max. :229.0 Max. :123.00
##       NA's :14  NA's :14
## txhtn   chol   hdl   triglyc   ldl
## No :1823 Min. : 95.0 Min. : 19.58 Min. : 25.0 Min. : 36.3
## Yes : 428 1st Qu.:189.0 1st Qu.: 42.00 1st Qu.: 72.0 1st Qu.:115.8
## NA's: 43 Median :215.0 Median : 51.00 Median : 97.0 Median :140.6
##    Mean :218.8 Mean : 52.69 Mean :115.6 Mean :143.2
##    3rd Qu.:245.0 3rd Qu.: 61.43 3rd Qu.:136.0 3rd Qu.:168.1
##    Max. :488.0 Max. :112.00 Max. :960.0 Max. :329.6
##    NA's :101  NA's :69  NA's :63  NA's :168
```

```
## histchol  txchol   height   weight   bmi
## Yes : 709 No :2011 Min. :137.0 Min. : 41.20 Min. :17.15
## No :1564 Yes : 228 1st Qu.:156.0 1st Qu.: 63.50 1st Qu.:24.38
## NA's: 21 NA's: 55 Median :162.5 Median : 73.00 Median :27.18
##       Mean :162.9 Mean : 73.44 Mean :27.64
##       3rd Qu.:169.2 3rd Qu.: 82.00 3rd Qu.:30.41
##       Max. :199.0 Max. :127.20 Max. :48.24
##       NA's :35  NA's :35  NA's :35
##  phyact   pcs    mcs   cv
## Min. : 0.0 Min. :13.95 Min. : 3.424 No :2071
## 1st Qu.: 159.5 1st Qu.:45.01 1st Qu.:42.181 Yes : 92
## Median : 303.7 Median :52.27 Median :51.260 NA's: 131
## Mean : 398.8 Mean :49.62 Mean :47.983
## 3rd Qu.: 521.9 3rd Qu.:55.78 3rd Qu.:56.047
## Max. :5083.2 Max. :67.06 Max. :69.904
## NA's :88  NA's :240  NA's :240
##  tocv   death   todeath
## Min. : 0.115 No :1975 Min. : 0.22
## 1st Qu.: 782.710 Yes : 173 1st Qu.: 787.63
## Median :1718.026 NA's: 146 Median :1668.40
## Mean :1754.668    Mean :1721.31
## 3rd Qu.:2690.549    3rd Qu.:2662.54
## Max. :3650.674    Max. :3651.26
## NA's :131     NA's :146
```

The data contains 2294 observations (rows) on the following 25 variables (columns):

1. **id:** Individual id
2. **year:** Recruitment year, a factor with levels 1995, 2000, 2005.
3. **age:** Patient age at recruitment date.
4. **sex:** Sex, a factor with levels male, female.
5. **smoker:** Smoking status, a factor with levels Never smoker,

Current or former < 1y, Never or former >= 1y.

6. **sbp:** Systolic blood pressure.
7. **dbp:** Diastolic blood pressure.
8. **histhtn:** History of hypertension, a factor with levels Yes, No.
9. **txhtn:** Hypertension (HTN) treatment, a factor with levels No, Yes.
10. **chol:** Total cholesterol (mg/dl).
11. **hdl:** HDL cholesterol (mg/dl).
12. **triglyc:** Triglycerides (mg/dl).
13. **ldl:** LDL cholesterol (mg/dl).
14. **histchol:** History of hypercholesterolemia, a factor with levels Yes, No.
15. **txchol:** Cholesterol treatment, a factor with levels No, Yes.
16. **height:** Height (cm).
17. **weight:** Weight (Kg).
18. **bmi:** Body mass index.
19. **phyact:** Physical activity (Kcal/week).
20. **pcs:** Physical component summary.
21. **mcs:** Mental component summary.
22. **death:** Overall death, a factor with levels No, Yes.
23. **todeath:** Days to overall death or end of follow-up.
24. **cv:** Cardiovascular event, a factor with levels No, Yes.
25. **tocv:** Days to cardiovascular event or end of follow-up.

We have 9 factor variables, year, sex, smoker, histhtn, txhtn, histchol, txchol, cv, and death.

by using table() function, we can get the counts for two or three variables at once.

example of crosstab for two variables.

```
table(regicor$sex,regicor$smoker)
##
##   Never smoker Current or former < 1y Former >= 1y
## Male   301     410   360
## Female   900     183   79
```

From the results, we see that:

1.301 males are never smokers compared to 410 males who are Current or former < 1y smokers and 360 males who are Former >= 1y smokers.

2.900 females are never smokers compared to 183 females who are Current or former < 1y smokers and 79 females who are Former >= 1y smokers.

```
table(regicor$smoker, regicor$histhtn)
##
##       Yes No
## Never smoker   421 777
## Current or former < 1y 125 464
## Former >= 1y   162 277
```

From the results, we see that:

1. 421 of the never smokers have a history of hypertension compared to 777 of such nonsmokers who do not have such history.
2. 125 of the Current or former < 1y smokers have a history of hypertension compared to 464 such smokers who do not have such history.
3. 162 of the Former >= 1y smokers have a history of hypertension compared to 277 such smokers who do not have such history.

### 6.2.3 Table() Function for Three Columns

```
table(regicor$sex,regicor$smoker,regicor$histhtn)
## , , = Yes
##
##
##  Never smoker Current or former < 1y Former >= 1y
## Male   85    101   145
## Female   336     24   17
##
## , , = No
##
```

```
##
## Never smoker Current or former < 1y Former >= 1y
## Male 214 306 215
## Female 563 158 62
```

From the results, we see that:

1. For those with a history of hypertension, there are 85 males and 336 females who are never smokers compared to 101 males and 24 females who are Current or former < 1y smokers and 145 males and 17 females who are Former >= 1y smokers.
2. For those without a history of hypertension, there are 214 males and 563 females who are never smokers compared to 306 males and 158 females who are Current or former < 1y smokers and 215 males and 62 females who are Former >= 1y smokers.

### 6.2.4 prop.table() Function

As looking at count numbers will not be meaningful to study the relation between variables, using the prop.table() function allows us to see the proportion of each category.

Using the prop.table() function to study the relation between sex and smoking. Using the prop.tab() alone will give the proportion of every cell. Multiplying the prop.table() function by 100, we will get the percentages of every cell instead of proportions.

```
prop.table(table(regicor$sex,regicor$smoker))
##
## Never smoker Current or former < 1y Former >= 1y
## Male 0.13479624 0.18360949 0.16121809
## Female 0.40304523 0.08195253 0.03537841
prop.table(table(regicor$sex,regicor$smoker))*100
##
## Never smoker Current or former < 1y Former >= 1y
## Male 13.479624 18.360949 16.121809
## Female 40.304523 8.195253 3.537841
```

From the result, we note that:

1. The sum of all these percentages is 100%.
2. Majority of our data is females nonsmokers (40.3%) followed by males Current or former < 1y smokers (18.4%), males Former >= 1y smokers (16.1%), males nonsmokers (13.5%), females Current or former < 1y smokers (8.2%).
3. Females Former >= 1y smokers are least represented in our data (3.5%).
4. This means, for example, that 40.3% of our rows (observations) are females nonsmokers and only 3.5% of our rows (observations) are females Former >= 1y smokers.

Using the prop.table() with 1 argument will generate proportions by rows.

```
prop.table(table(regicor$sex,regicor$smoker),1)
##
##   Never smoker Current or former < 1y Former >= 1y
## Male  0.28104575    0.38281979 0.33613445
## Female 0.77452668    0.15748709 0.06798623
prop.table(table(regicor$sex,regicor$smoker),1)*100
##
##   Never smoker Current or former < 1y Former >= 1y
## Male  28.104575    38.281979 33.613445
## Female 77.452668    15.748709  6.798623
```

From the results, we note that:

1. The sum of percentages for each row, males or females, is 100%.
2. 28.1% of males are nonsmokers compared to 77.5% of females.
3. 38.3% of males are Current or former < 1y smokers compared to 15.7% of females.
4. 33.6% of males are Former >= 1y smokers compared to only 6.8% of females.
5. From these results, it is shown that there is a gender difference in different smoking habits.

Using the prop.table() with 2 argument will generate proportions by columns.

```
prop.table(table(regicor$sex,regicor$smoker),2)
##
## Never smoker Current or former < 1y Former >= 1y
## Male 0.2506245 0.6913997 0.8200456
## Female 0.7493755 0.3086003 0.1799544
prop.table(table(regicor$sex,regicor$smoker),2)*100
##
## Never smoker Current or former < 1y Former >= 1y
## Male 25.06245 69.13997 82.00456
## Female 74.93755 30.86003 17.99544
```

From the results, we note that:

1. The sum of percentages for each column, Never smoker, Current or former < 1y, or Former >= 1y, is 100%.
2. 25.1% of nonsmokers are males compared to 74.9% of nonsmokers are females.
3. 69.1% of Current or former < 1y smokers are males compared to 30.9% of them are females.
4. 82% of Former >= 1y smokers are males compared to only 18% of them are females.
5. From these results, it is shown that there is a gender difference in different smoking habits.

## 6.3 CUT() FUNCTION

cut() function is an important function in converting numeric vectors to factors or categories. This is an important step in many data analysis tasks.

cut() function has an argument, breaks, that is a vector of unique values that defines the break points. Breaks can also be a single number to define the number of intervals of equal length.

By default, the intervals are created so that the lowest number of range is not included (it is included in previous range) and largest number of range is included.

### 6.3.1 Cut() Function with Breaks as a Vector of Break Points

```
x<–1:100

x
## [1] 1 2 3 4 5 6 7 8 9 10 11 12 13 14 15 16 17 18
## [19] 19 20 21 22 23 24 25 26 27 28 29 30 31 32 33 34 35 36
## [37] 37 38 39 40 41 42 43 44 45 46 47 48 49 50 51 52 53 54
## [55] 55 56 57 58 59 60 61 62 63 64 65 66 67 68 69 70 71 72
## [73] 73 74 75 76 77 78 79 80 81 82 83 84 85 86 87 88 89 90
## [91] 91 92 93 94 95 96 97 98 99 100
y<-cut(x, breaks = c(1,50,100))

y
## [1] <NA> (1,50] (1,50] (1,50] (1,50] (1,50] (1,50] (1,50]
## [9] (1,50] (1,50] (1,50] (1,50] (1,50] (1,50] (1,50] (1,50]
## [17] (1,50] (1,50] (1,50] (1,50] (1,50] (1,50] (1,50] (1,50]
## [25] (1,50] (1,50] (1,50] (1,50] (1,50] (1,50] (1,50] (1,50]
## [33] (1,50] (1,50] (1,50] (1,50] (1,50] (1,50] (1,50] (1,50]
## [41] (1,50] (1,50] (1,50] (1,50] (1,50] (1,50] (1,50] (1,50]
## [49] (1,50] (1,50] (50,100] (50,100] (50,100] (50,100] (50,100] (50,100]
## [57] (50,100] (50,100] (50,100] (50,100] (50,100] (50,100] (50,100]
(50,100]
## [65] (50,100] (50,100] (50,100] (50,100] (50,100] (50,100] (50,100]
(50,100]
## [73] (50,100] (50,100] (50,100] (50,100] (50,100] (50,100] (50,100]
(50,100]
## [81] (50,100] (50,100] (50,100] (50,100] (50,100] (50,100] (50,100]
(50,100]
## [89] (50,100] (50,100] (50,100] (50,100] (50,100] (50,100] (50,100]
(50,100]
## [97] (50,100] (50,100] (50,100] (50,100]
```

```
## Levels: (1,50] (50,100]
table(y)
## y
## (1,50] (50,100]
##  49 50
y<-cut(x, breaks = c(1,50,100), include.lowest = TRUE)


y
## [1] [1,50] [1,50] [1,50] [1,50] [1,50] [1,50] [1,50] [1,50]
## [9] [1,50] [1,50] [1,50] [1,50] [1,50] [1,50] [1,50] [1,50]
## [17] [1,50] [1,50] [1,50] [1,50] [1,50] [1,50] [1,50] [1,50]
## [25] [1,50] [1,50] [1,50] [1,50] [1,50] [1,50] [1,50] [1,50]
## [33] [1,50] [1,50] [1,50] [1,50] [1,50] [1,50] [1,50] [1,50]
## [41] [1,50] [1,50] [1,50] [1,50] [1,50] [1,50] [1,50] [1,50]
## [49] [1,50] [1,50] (50,100] (50,100] (50,100] (50,100] (50,100] (50,100]
## [57] (50,100] (50,100] (50,100] (50,100] (50,100] (50,100] (50,100]
(50,100]
## [65] (50,100] (50,100] (50,100] (50,100] (50,100] (50,100] (50,100]
(50,100]
## [73] (50,100] (50,100] (50,100] (50,100] (50,100] (50,100] (50,100]
(50,100]
## [81] (50,100] (50,100] (50,100] (50,100] (50,100] (50,100] (50,100]
(50,100]
## [89] (50,100] (50,100] (50,100] (50,100] (50,100] (50,100] (50,100]
(50,100]
## [97] (50,100] (50,100] (50,100] (50,100]
## Levels: [1,50] (50,100]
table(y)
## y
## [1,50] (50,100]
##  50 50
y<-cut(x, breaks = c(0,50,100))
```

```
y
## [1] (0,50] (0,50] (0,50] (0,50] (0,50] (0,50] (0,50] (0,50]
## [9] (0,50] (0,50] (0,50] (0,50] (0,50] (0,50] (0,50] (0,50]
## [17] (0,50] (0,50] (0,50] (0,50] (0,50] (0,50] (0,50] (0,50]
## [25] (0,50] (0,50] (0,50] (0,50] (0,50] (0,50] (0,50] (0,50]
## [33] (0,50] (0,50] (0,50] (0,50] (0,50] (0,50] (0,50] (0,50]
## [41] (0,50] (0,50] (0,50] (0,50] (0,50] (0,50] (0,50] (0,50]
## [49] (0,50] (0,50] (50,100] (50,100] (50,100] (50,100] (50,100] (50,100]
## [57] (50,100] (50,100] (50,100] (50,100] (50,100] (50,100] (50,100]
(50,100]
## [65] (50,100] (50,100] (50,100] (50,100] (50,100] (50,100] (50,100]
(50,100]
## [73] (50,100] (50,100] (50,100] (50,100] (50,100] (50,100] (50,100]
(50,100]
## [81] (50,100] (50,100] (50,100] (50,100] (50,100] (50,100] (50,100]
(50,100]
## [89] (50,100] (50,100] (50,100] (50,100] (50,100] (50,100] (50,100]
(50,100]
## [97] (50,100] (50,100] (50,100] (50,100]
## Levels: (0,50] (50,100]
table(y)
## y
## (0,50] (50,100]
##  50  50
# we can define any range

y<-cut(x, breaks = c(0,40,100))

y
## [1] (0,40] (0,40] (0,40] (0,40] (0,40] (0,40] (0,40] (0,40]
## [9] (0,40] (0,40] (0,40] (0,40] (0,40] (0,40] (0,40] (0,40]
```

```
## [17] (0,40] (0,40] (0,40] (0,40] (0,40] (0,40] (0,40] (0,40]
## [25] (0,40] (0,40] (0,40] (0,40] (0,40] (0,40] (0,40] (0,40]
## [33] (0,40] (0,40] (0,40] (0,40] (0,40] (0,40] (0,40] (0,40]
## [41] (40,100] (40,100] (40,100] (40,100] (40,100] (40,100] (40,100]
(40,100]
## [49] (40,100] (40,100] (40,100] (40,100] (40,100] (40,100] (40,100]
(40,100]
## [57] (40,100] (40,100] (40,100] (40,100] (40,100] (40,100] (40,100]
(40,100]
## [65] (40,100] (40,100] (40,100] (40,100] (40,100] (40,100] (40,100]
(40,100]
## [73] (40,100] (40,100] (40,100] (40,100] (40,100] (40,100] (40,100]
(40,100]
## [81] (40,100] (40,100] (40,100] (40,100] (40,100] (40,100] (40,100]
(40,100]
## [89] (40,100] (40,100] (40,100] (40,100] (40,100] (40,100] (40,100]
(40,100]
## [97] (40,100] (40,100] (40,100] (40,100]
## Levels: (0,40] (40,100]
table(y)
## y
## (0,40] (40,100]
##  40  60
```

Using cut function with 3 break points (1,50,100) means that I need to create two ranges or categories,1–50 and 50–100. However, defining the ranges as 1–50 and 50–100 will drop out the first value (1) because it is not included by default in R. This is evident from the table() output function where there are 49 values in first category and 50 values in second category. The label, (1,50], indicates that 1 is not included in that interval and 50 is included in that interval.

By adding the argument, include.lowest = TRUE, will correct this problem as evident from the table() output function. There are 50 values in first range or category and 50 other values in second range or category. The label, [1,50], now indicates that 1 and 50 are included in that interval.

Another option is defining break points as 0,50,100 (0 is lower than 1 so 1 will be included in first category) will also correct this problem as evident from the table() output function.

### 6.3.2 Cut() Function with Breaks as a Single Number

```
x<-1:100

y<-cut(x, breaks = 2)

y
## [1] (0.901,50.5] (0.901,50.5] (0.901,50.5] (0.901,50.5] (0.901,50.5]
## [6] (0.901,50.5] (0.901,50.5] (0.901,50.5] (0.901,50.5] (0.901,50.5]
## [11] (0.901,50.5] (0.901,50.5] (0.901,50.5] (0.901,50.5] (0.901,50.5]
## [16] (0.901,50.5] (0.901,50.5] (0.901,50.5] (0.901,50.5] (0.901,50.5]
## [21] (0.901,50.5] (0.901,50.5] (0.901,50.5] (0.901,50.5] (0.901,50.5]
## [26] (0.901,50.5] (0.901,50.5] (0.901,50.5] (0.901,50.5] (0.901,50.5]
## [31] (0.901,50.5] (0.901,50.5] (0.901,50.5] (0.901,50.5] (0.901,50.5]
## [36] (0.901,50.5] (0.901,50.5] (0.901,50.5] (0.901,50.5] (0.901,50.5]
## [41] (0.901,50.5] (0.901,50.5] (0.901,50.5] (0.901,50.5] (0.901,50.5]
## [46] (0.901,50.5] (0.901,50.5] (0.901,50.5] (0.901,50.5] (0.901,50.5]
## [51] (50.5,100] (50.5,100] (50.5,100] (50.5,100] (50.5,100]
## [56] (50.5,100] (50.5,100] (50.5,100] (50.5,100] (50.5,100]
## [61] (50.5,100] (50.5,100] (50.5,100] (50.5,100] (50.5,100]
## [66] (50.5,100] (50.5,100] (50.5,100] (50.5,100] (50.5,100]
## [71] (50.5,100] (50.5,100] (50.5,100] (50.5,100] (50.5,100]
## [76] (50.5,100] (50.5,100] (50.5,100] (50.5,100] (50.5,100]
## [81] (50.5,100] (50.5,100] (50.5,100] (50.5,100] (50.5,100]
## [86] (50.5,100] (50.5,100] (50.5,100] (50.5,100] (50.5,100]
## [91] (50.5,100] (50.5,100] (50.5,100] (50.5,100] (50.5,100]
## [96] (50.5,100] (50.5,100] (50.5,100] (50.5,100] (50.5,100]
## Levels: (0.901,50.5] (50.5,100]
```

```
table(y)
## y
## (0.901,50.5] (50.5,100]
##   50   50
y<-cut(x, breaks = 3)

y
## [1] (0.901,34] (0.901,34] (0.901,34] (0.901,34] (0.901,34] (0.901,34]
## [7] (0.901,34] (0.901,34] (0.901,34] (0.901,34] (0.901,34] (0.901,34]
## [13] (0.901,34] (0.901,34] (0.901,34] (0.901,34] (0.901,34] (0.901,34]
## [19] (0.901,34] (0.901,34] (0.901,34] (0.901,34] (0.901,34] (0.901,34]
## [25] (0.901,34] (0.901,34] (0.901,34] (0.901,34] (0.901,34] (0.901,34]
## [31] (0.901,34] (0.901,34] (0.901,34] (0.901,34] (34,67] (34,67]
## [37] (34,67] (34,67] (34,67] (34,67] (34,67] (34,67]
## [43] (34,67] (34,67] (34,67] (34,67] (34,67] (34,67]
## [49] (34,67] (34,67] (34,67] (34,67] (34,67] (34,67]
## [55] (34,67] (34,67] (34,67] (34,67] (34,67] (34,67]
## [61] (34,67] (34,67] (34,67] (34,67] (34,67] (34,67]
## [67] (34,67] (67,100] (67,100] (67,100] (67,100] (67,100]
## [73] (67,100] (67,100] (67,100] (67,100] (67,100] (67,100]
## [79] (67,100] (67,100] (67,100] (67,100] (67,100] (67,100]
## [85] (67,100] (67,100] (67,100] (67,100] (67,100] (67,100]
## [91] (67,100] (67,100] (67,100] (67,100] (67,100] (67,100]
## [97] (67,100] (67,100] (67,100] (67,100]
## Levels: (0.901,34] (34,67] (67,100]
table(y)
## y
## (0.901,34] (34,67] (67,100]
##   34   33   33
y<-cut(x, breaks = 11)
```

```
y
## [1] (0.901,10] (0.901,10] (0.901,10] (0.901,10] (0.901,10] (0.901,10]
## [7] (0.901,10] (0.901,10] (0.901,10] (0.901,10] (10,19] (10,19]
## [13] (10,19] (10,19] (10,19] (10,19] (10,19] (10,19]
## [19] (10,19] (19,28] (19,28] (19,28] (19,28] (19,28]
## [25] (19,28] (19,28] (19,28] (19,28] (28,37] (28,37]
## [31] (28,37] (28,37] (28,37] (28,37] (28,37] (28,37]
## [37] (28,37] (37,46] (37,46] (37,46] (37,46] (37,46]
## [43] (37,46] (37,46] (37,46] (37,46] (46,55] (46,55]
## [49] (46,55] (46,55] (46,55] (46,55] (46,55] (46,55]
## [55] (46,55] (55,64] (55,64] (55,64] (55,64] (55,64]
## [61] (55,64] (55,64] (55,64] (55,64] (64,73] (64,73]
## [67] (64,73] (64,73] (64,73] (64,73] (64,73] (64,73]
## [73] (64,73] (73,82] (73,82] (73,82] (73,82] (73,82]
## [79] (73,82] (73,82] (73,82] (73,82] (82,91] (82,91]
## [85] (82,91] (82,91] (82,91] (82,91] (82,91] (82,91]
## [91] (82,91] (91,100] (91,100] (91,100] (91,100] (91,100]
## [97] (91,100] (91,100] (91,100] (91,100]
## 11 Levels: (0.901,10] (10,19] (19,28] (28,37] (37,46] (46,55] ... (91,100]
table(y)
## y
## (0.901,10] (10,19] (19,28] (28,37] (37,46] (46,55] (55,64]
##   10 9 9 9 9 9 9
## (64,73] (73,82] (82,91] (91,100]
##   9 9 9 9
```

Using the breaks with single number will create ranges of equal length. for example, 100–50.5 = 49.5 = 50.5–0.901. Note also that first range of 2 categories is 0.901–50.5 to include both 1 and 50 in that range.

Note also that the same rule applies to round and square brackets. For example, range (0.901,34], means that 0.901 value is not included in that range but values larger than 0.901 till 34 are included in that range.

### 6.3.3 Cut() Function with Labels

Using the labels argument allows you to set meaningful names to your ranges or categories.

```
x<–1:100

y<-cut(x, breaks = 2, labels = c("first_half","second_half"))

y
## [1] first_half first_half first_half first_half first_half first_half
## [7] first_half first_half first_half first_half first_half first_half
## [13] first_half first_half first_half first_half first_half first_half
## [19] first_half first_half first_half first_half first_half first_half
## [25] first_half first_half first_half first_half first_half first_half
## [31] first_half first_half first_half first_half first_half first_half
## [37] first_half first_half first_half first_half first_half first_half
## [43] first_half first_half first_half first_half first_half first_half
## [49] first_half first_half second_half second_half second_half second_
half
## [55] second_half second_half second_half second_half second_half
second_half
## [61] second_half second_half second_half second_half second_half
second_half
## [67] second_half second_half second_half second_half second_half
second_half
## [73] second_half second_half second_half second_half second_half
second_half
## [79] second_half second_half second_half second_half second_half
second_half
## [85] second_half second_half second_half second_half second_half
```

```
second_half
## [91] second_half second_half second_half second_half second_half
second_half
## [97] second_half second_half second_half second_half
## Levels: first_half second_half
table(y)
## y
## first_half second_half
##   50   50
```

Example of cut() function with labels for the regicor data.

```
library(compareGroups)

data("regicor")

summary(regicor$age)
## Min. 1st Qu. Median Mean 3rd Qu. Max.
## 35.00 46.00 55.00 54.74 64.00 74.00
regicor$age_category<-cut(regicor$age, breaks = c(34,40,60,74),

    labels = c("young","old","very old"))

str(regicor)
## 'data.frame': 2294 obs. of 26 variables:
## $ id   : num 2.26e+03 1.88e+03 3.00e+09 3.00e+09 3.00e+09 ...
## ..- attr(*, "label")= Named chr "Individual id"
## .. ..- attr(*, "names")= chr "id"
## $ year  : Factor w/ 3 levels "1995","2000",..: 3 3 2 2 2 2 2 1 3 1 ...
## ..- attr(*, "label")= Named chr "Recruitment year"
## .. ..- attr(*, "names")= chr "year"
```

```
## $ age   : int 70 56 37 69 70 40 66 53 43 70 ...
## ..- attr(*, “label”)= Named chr “Age”
## .. ..- attr(*, “names”)= chr “age”
## $ sex   : Factor w/ 2 levels “Male”,”Female”: 2 2 1 2 2 2 1 2 2 1 ...
## ..- attr(*, “label”)= chr “Sex”
## $ smoker  : Factor w/ 3 levels “Never smoker”,..: 1 1 2 1 NA 2 1 1 3 3 ...
## ..- attr(*, “label”)= Named chr “Smoking status”
## .. ..- attr(*, “names”)= chr “smoker”
## $ sbp   : int 138 139 132 168 NA 108 120 132 95 142 ...
## ..- attr(*, “label”)= Named chr “Systolic blood pressure”
## .. ..- attr(*, “names”)= chr “sbp”
## $ dbp   : int 75 89 82 97 NA 70 72 78 65 78 ...
## ..- attr(*, “label”)= Named chr “Diastolic blood pressure”
## .. ..- attr(*, “names”)= chr “dbp”
## $ histhtn  : Factor w/ 2 levels “Yes”,”No”: 2 2 2 2 2 2 1 2 2 2 ...
## ..- attr(*, “label”)= Named chr “History of hypertension”
## .. ..- attr(*, “names”)= chr “histbp”
## $ txhtn  : Factor w/ 2 levels “No”,”Yes”: 1 1 1 1 1 1 2 1 1 1 ...
## ..- attr(*, “label”)= chr “Hypertension treatment”
## $ chol  : num 294 220 245 168 NA NA 298 254 194 188 ...
## ..- attr(*, “label”)= Named chr “Total cholesterol”
## .. ..- attr(*, “names”)= chr “chol”
## $ hdl   : num 57 50 59.8 53.2 NA ...
## ..- attr(*, “label”)= Named chr “HDL cholesterol”
## .. ..- attr(*, “names”)= chr “hdl”
## $ triglyc  : num 93 160 89 116 NA 94 71 NA 68 137 ...
## ..- attr(*, “label”)= Named chr “Triglycerides”
## .. ..- attr(*, “names”)= chr “triglyc”
## $ ldl   : num 218.4 138 167.4 91.6 NA ...
## ..- attr(*, “label”)= Named chr “LDL cholesterol”
## .. ..- attr(*, “names”)= chr “ldl”
```

```
## $ histchol : Factor w/ 2 levels "Yes","No": 2 2 2 2 NA 2 1 2 2 2 ...
## ..- attr(*, "label")= chr "History of hyperchol."
## $ txchol  : Factor w/ 2 levels "No","Yes": 1 1 1 1 NA 1 1 1 1 1 ...
## ..- attr(*, "label")= Named chr "Cholesterol treatment"
## .. ..- attr(*, "names")= chr "txchol"
## $ height  : num 160 163 170 147 NA ...
## ..- attr(*, "label")= Named chr "Height (cm)"
## .. ..- attr(*, "names")= chr "height"
## $ weight  : num 64 67 70 68 NA 43.5 79.2 45.8 53 62 ...
## ..- attr(*, "label")= Named chr "Weight (Kg)"
## .. ..- attr(*, "names")= chr "weight"
## $ bmi   : num 25 25.2 24.2 31.5 NA ...
## ..- attr(*, "label")= Named chr "Body mass index"
## .. ..- attr(*, "names")= chr "bmi"
## $ phyact  : num 304 160 553 522 NA ...
## ..- attr(*, "label")= Named chr "Physical activity (Kcal/week)"
## .. ..- attr(*, "names")= chr "phyact"
## $ pcs   : num 54.5 58.2 43.4 54.3 NA ...
## ..- attr(*, "label")= Named chr "Physical component"
## .. ..- attr(*, "names")= chr "pcs"
## $ mcs   : num 58.9 48 62.6 57.9 NA ...
## ..- attr(*, "label")= chr "Mental component"
## $ cv   : Factor w/ 2 levels "No","Yes": 1 1 1 1 NA 1 1 1 1 1 ...
## ..- attr(*, "label")= chr "Cardiovascular event"
## $ tocv  : num 1025 2757 1906 1055 NA ...
## ..- attr(*, "label")= chr "Days to cardiovascular event or end of follow-
up"
## $ death  : Factor w/ 2 levels "No","Yes": 2 1 1 1 NA 1 2 1 1 1 ...
## ..- attr(*, "label")= chr "Overall death"
## $ todeath  : num 1299.2 39.3 858.4 1833.1 NA ...
## ..- attr(*, "label")= chr "Days to overall death or end of follow-up"
```

```
## $ age_category: Factor w/ 3 levels "young","old",..: 3 2 1 3 3 1 3 2 2 3 ...
head(regicor)
##    id year age sex     smoker sbp dbp histhtn txhtn
## 6101  2265 2005 70 Female   Never smoker 138 75  No No
## 5762  1882 2005 56 Female   Never smoker 139 89  No No
## 2992 3000105616 2000 37 Male Current or former < 1y 132 82  No No
## 2611 3000103485 2000 69 Female   Never smoker 168 97  No No
## 2762 3000103963 2000 70 Female     <NA> NA NA  No No
## 1516 3000100883 2000 40 Female Current or former < 1y 108 70  No No
##  chol  hdl triglyc  ldl histchol txchol height weight  bmi
## 6101 294 57.00000  93 218.40000  No  No 160 64.0 25.00000
## 5762 220 50.00000  160 138.00000  No  No 163 67.0 25.21736
## 2992 245 59.80429  89 167.39571  No  No 170 70.0 24.22145
## 2611 168 53.17571  116 91.62429  No  No 147 68.0 31.46837
## 2762 NA  NA  NA  NA  <NA> <NA>  NA  NA  NA
## 1516 NA 68.90000  94  NA  No  No 158 43.5 17.42509
##  phyact pcs mcs cv  tocv death todeath age_category
## 6101 304.2000 54.455 58.918 No 1024.882 Yes 1299.16343  very old
## 5762 160.3000 58.165 47.995 No 2756.849 No 39.32629   old
## 2992 552.7912 43.429 62.585 No 1905.969 No 858.42203  young
## 2611 522.0000 54.325 57.900 No 1055.380 No 1833.07619  very old
## 2762  NA  NA  NA <NA>  NA <NA>   NA  very old
## 1516 386.9505 57.315 47.869 No 3239.241 No 877.61155  young
table(regicor$age_category)
##
## young  old very old
##  285  1242  767
table(regicor$age_category, regicor$age)
##
##   35 36 37 38 39 40 41 42 43 44 45 46 47 48 49 50 51 52 53 54 55 56 57
## young 41 41 44 55 48 56 0 0 0 0 0 0 0 0 0 0 0 0 0 0 0 0 0
```

```
## old  0 0 0 0 0 0 49 58 59 52 56 70 72 70 63 68 49 52 75 60 48 64 68
## very old 0 0 0 0 0 0 0 0 0 0 0 0 0 0 0 0 0 0 0 0 0 0 0
##
##   58 59 60 61 62 63 64 65 66 67 68 69 70 71 72 73 74
## young  0 0 0 0 0 0 0 0 0 0 0 0 0 0 0 0 0
## old  81 67 61 0 0 0 0 0 0 0 0 0 0 0 0 0 0
## very old 0 0 0 53 50 57 54 48 50 68 65 52 68 64 40 46 52
```

Regicor data has a column for participant age. By using the summary function, we noted that the age has a range from 35 to 74 years. we wish to cut this column into 3 ranges or categories, young (34–40), old (41–60), and very old (61–74). Therefore, we define cut the points at 34,40,60,74. We choose 34 to include 35 (minimum value) in the first range. We stored the result in a new created column called “age_category”.

The str() and head() functions show that a new column named “age_category” has been added to our dataframe. The dataframe is now composed of 26 columns instead of 25 columns.

Using the table() function with this new column showed there are 285 rows for young category, 1242 rows for old category, and 767 rows for very old category. The total is 2294 which is the number of rows for the regicor dataframe.

Using the table() function with the age-categorized column and the age column, before categorization, showed that the categorization was done correctly. For example, the young category contains 41, 41, 44, 55, 48, and 56 rows for ages 35, 36, 37, 38, 39, and 40, respectively. The young category contains zero values for all other age numbers.

## 6.4. SPLIT() FUNCTION

split() function divides the data in the vector x into the groups defined by a factor, f. f is recycled as necessary and if the length of x is not a multiple of the length of f a warning is printed. The value returned from split is a list of vectors containing the values for the groups. The components of the list are named by the levels of f.

### 6.4.1. Split() for Vectors

Example:

```
#define the vector

x<-1:10

x
## [1] 1 2 3 4 5 6 7 8 9 10
#define a factor of two levels

f<-factor(c(0,1))

f
## [1] 0 1
## Levels: 0 1
# use split function to produce the splitted list

l<-split(x,f)

l
## $`0`
## [1] 1 3 5 7 9
##
## $`1`
## [1] 2 4 6 8 10
```

Because the factor is of only two levels so it is recycled.

```
#define the vector

x<-1:10

x
```

```
## [1] 1 2 3 4 5 6 7 8 9 10
#define a factor of three levels

f<-factor(c(0,0,0,0,1,1,1,2,2,2))

f
## [1] 0 0 0 0 1 1 1 2 2 2
## Levels: 0 1 2
# use split function to produce the splitted list

l<-split(x,f)

l
## $`0`
## [1] 1 2 3 4
##
## $`1`
## [1] 5 6 7
##
## $`2`
## [1] 8 9 10
```

By defining the factor levels, we can cut the vector into any number of pieces.

### *6.4.2. Split() for Dataframes*

Usually, the split() function is used for dataframes or its numerical columns to get powerful summaries from the splitted data for the relation between numerical and categorical columns.

Using the regicor dataframe, we will use the split() function to get the mean age, weight, and height (numerical columns) of both sexes (categorical column).

```
table(regicor$sex)
##
## Male Female
## 1101 1193
l<-split(regicor,regicor$sex)

str(l)
## List of 2
## $ Male :'data.frame': 1101 obs. of 26 variables:
## ..$ id   : num [1:1101] 3.00e+09 3.00e+09 1.00e+09 2.26e+03 3.18e+03
...
## ..$ year  : Factor w/ 3 levels "1995","2000",..: 2 2 1 3 3 3 3 3 1 1 ...
## ..$ age   : int [1:1101] 37 66 70 54 54 68 42 70 70 56 ...
## ..$ sex   : Factor w/ 2 levels "Male","Female": 1 1 1 1 1 1 1 1 1 1 ...
## ..$ smoker  : Factor w/ 3 levels "Never smoker",..: 2 1 3 3 1 3 2 3 1 3 ...
## ..$ sbp   : int [1:1101] 132 120 142 130 117 158 128 195 132 124 ...
## ..$ dbp   : int [1:1101] 82 72 78 66 70 71 84 97 74 82 ...
## ..$ histhtn  : Factor w/ 2 levels "Yes","No": 2 1 2 2 2 1 1 1 2 2 ...
## ..$ txhtn  : Factor w/ 2 levels "No","Yes": 1 2 1 1 1 2 1 2 1 1 ...
## ..$ chol  : num [1:1101] 245 298 188 268 211 209 218 182 183 254 ...
## ..$ hdl   : num [1:1101] 59.8 78.9 35.8 37 49 ...
## ..$ triglyc  : num [1:1101] 89 71 137 128 144 150 372 113 95 145 ...
## ..$ ldl   : num [1:1101] 167 205 125 205 133 ...
## ..$ histchol : Factor w/ 2 levels "Yes","No": 2 1 2 2 1 2 2 2 2 1 ...
## ..$ txchol  : Factor w/ 2 levels "No","Yes": 1 1 1 1 1 1 1 1 1 1 ...
## ..$ height  : num [1:1101] 170 164 160 163 167 156 180 156 162 155 ...
## ..$ weight  : num [1:1101] 70 79.2 62 79 65 69 88 80 77 72 ...
## ..$ bmi   : num [1:1101] 24.2 29.4 24.2 29.7 23.3 ...
## ..$ phyact  : num [1:1101] 553 478 208 285 208 ...
## ..$ pcs   : num [1:1101] 43.4 59.6 44 52 54.5 ...
```

```
## ..$ mcs   : num [1:1101] 62.6 40.8 58.6 38.2 58.9 ...
## ..$ cv   : Factor w/ 2 levels "No","Yes": 1 1 1 1 1 1 1 1 1 1 ...
## ..$ tocv  : num [1:1101] 1906 63.5 244.6 3054.9 2395.6 ...
## ..$ death  : Factor w/ 2 levels "No","Yes": 1 2 1 1 1 1 1 1 1 1 ...
## ..$ todeath  : num [1:1101] 858 2040 2029 2856 2468 ...
## ..$ age_category: Factor w/ 3 levels "young","old",..: 1 3 3 2 2 3 2 3 3 2
...
## $ Female:'data.frame': 1193 obs. of 26 variables:
## ..$ id   : num [1:1193] 2.26e+03 1.88e+03 3.00e+09 3.00e+09 3.00e+09
...
## ..$ year  : Factor w/ 3 levels "1995","2000",..: 3 3 2 2 2 1 3 3 3 3 ...
## ..$ age   : int [1:1193] 70 56 69 70 40 53 43 42 48 48 ...
## ..$ sex   : Factor w/ 2 levels "Male","Female": 2 2 2 2 2 2 2 2 2 2 ...
## ..$ smoker  : Factor w/ 3 levels "Never smoker",..: 1 1 1 NA 2 1 3 2 2 1 ...
## ..$ sbp   : int [1:1193] 138 139 168 NA 108 132 95 99 105 114 ...
## ..$ dbp   : int [1:1193] 75 89 97 NA 70 78 65 60 73 81 ...
## ..$ histhtn  : Factor w/ 2 levels "Yes","No": 2 2 2 2 2 2 2 1 2 1 ...
## ..$ txhtn  : Factor w/ 2 levels "No","Yes": 1 1 1 1 1 1 1 1 1 2 ...
## ..$ chol  : num [1:1193] 294 220 168 NA NA 254 194 116 162 163 ...
## ..$ hdl  : num [1:1193] 57 50 53.2 NA 68.9 ...
## ..$ triglyc  : num [1:1193] 93 160 116 NA 94 NA 68 39 45 63 ...
## ..$ ldl   : num [1:1193] 218.4 138 91.6 NA NA ...
## ..$ histchol : Factor w/ 2 levels "Yes","No": 2 2 2 NA 2 2 2 2 2 2 ...
## ..$ txchol  : Factor w/ 2 levels "No","Yes": 1 1 1 NA 1 1 1 1 1 1 ...
## ..$ height  : num [1:1193] 160 163 147 NA 158 ...
## ..$ weight  : num [1:1193] 64 67 68 NA 43.5 45.8 53 66 60 69 ...
## ..$ bmi   : num [1:1193] 25 25.2 31.5 NA 17.4 ...
## ..$ phyact  : num [1:1193] 304 160 522 NA 387 ...
## ..$ pcs   : num [1:1193] 54.5 58.2 54.3 NA 57.3 ...
## ..$ mcs   : num [1:1193] 58.9 48 57.9 NA 47.9 ...
## ..$ cv   : Factor w/ 2 levels "No","Yes": 1 1 1 NA 1 1 1 1 2 1 ...
## ..$ tocv  : num [1:1193] 1025 2757 1055 NA 3239 ...
```

```
## ..$ death  : Factor w/ 2 levels "No","Yes": 2 1 1 NA 1 1 1 1 1 1 ...
## ..$ todeath  : num [1:1193] 1299.2 39.3 1833.1 NA 877.6 ...
## ..$ age_category: Factor w/ 3 levels "young","old",..: 3 2 3 3 1 2 2 2 2 2
...
#mean age

mean(l$Male$age)
## [1] 54.78474
mean(l$Female$age)
## [1] 54.69153
# mean height

mean(l$Male$height)
## [1] NA
mean(l$Male$height, na.rm = TRUE)
## [1] 169.2727
mean(l$Female$height)
## [1] NA
mean(l$Female$height, na.rm = TRUE)
## [1] 156.9882
# mean weight

mean(l$Male$weight)
## [1] NA
mean(l$Male$weight, na.rm = TRUE)
## [1] 79.71587
mean(l$Female$weight)
## [1] NA
mean(l$Female$weight, na.rm = TRUE)
## [1] 67.58024
```

The sex column has splitted the dataframe to two dataframes, one for males with 1101 rows and the other for females with 1193 rows. These are the same number of rows produced by the table() function.

The mean age, height, and weight for males is 54.78, 169.27, and 79.72, respectively. The mean age, height, and weight for females is 54.69, 156.99, and 67.58, respectively.

We add the argument, na.rm = TRUE, to remove NA (missing) values from the height and weight values. Otherwise, NA will be produced as a result of the mean() function for values containing NA values.

Mean diastolic and systolic blood pressure by smoking habit.

```
table(regicor$smoker)
##
## Never smoker Current or former < 1y Former >= 1y
## 1201 593 439
l<-split(regicor,regicor$smoker)

str(l)
## List of 3
## $ Never smoker :'data.frame': 1201 obs. of 26 variables:
## ..$ id : num [1:1201] 2.26e+03 1.88e+03 3.00e+09 3.00e+09 1.00e+09
...
## ..$ year : Factor w/ 3 levels "1995","2000",..: 3 3 2 2 1 3 3 2 2 1 ...
## ..$ age : int [1:1201] 70 56 69 66 53 54 48 35 72 70 ...
## ..$ sex : Factor w/ 2 levels "Male","Female": 2 2 2 1 2 1 2 2 2 1 ...
## ..$ smoker : Factor w/ 3 levels "Never smoker",..: 1 1 1 1 1 1 1 1 1 1 ...
## ..$ sbp : int [1:1201] 138 139 168 120 132 117 114 111 155 132 ...
## ..$ dbp : int [1:1201] 75 89 97 72 78 70 81 70 80 74 ...
## ..$ histhtn : Factor w/ 2 levels "Yes","No": 2 2 2 1 2 2 1 2 1 2 ...
## ..$ txhtn : Factor w/ 2 levels "No","Yes": 1 1 1 2 1 1 2 1 1 1 ...
## ..$ chol : num [1:1201] 294 220 168 298 254 211 163 238 NA 183 ...
## ..$ hdl : num [1:1201] 57 50 53.2 78.9 86 ...
## ..$ triglyc : num [1:1201] 93 160 116 71 NA 144 63 108 106 95 ...
```

```
## ..$ ldl   : num [1:1201] 218.4 138 91.6 204.9 NA ...
## ..$ histchol : Factor w/ 2 levels "Yes","No": 2 2 2 1 2 1 2 1 1 2 ...
## ..$ txchol  : Factor w/ 2 levels "No","Yes": 1 1 1 1 1 1 1 1 1 1 ...
## ..$ height  : num [1:1201] 160 163 147 164 154 ...
## ..$ weight  : num [1:1201] 64 67 68 79.2 45.8 65 69 76 69 77 ...
## ..$ bmi   : num [1:1201] 25 25.2 31.5 29.4 19.2 ...
## ..$ phyact  : num [1:1201] 304 160 522 478 324 ...
## ..$ pcs   : num [1:1201] 54.5 58.2 54.3 59.6 49.5 ...
## ..$ mcs   : num [1:1201] 58.9 48 57.9 40.8 57.4 ...
## ..$ cv   : Factor w/ 2 levels "No","Yes": 1 1 1 1 1 1 1 1 1 1 ...
## ..$ tocv  : num [1:1201] 1024.9 2756.8 1055.4 63.5 1974.1 ...
## ..$ death  : Factor w/ 2 levels "No","Yes": 2 1 1 2 1 1 1 1 1 1 ...
## ..$ todeath  : num [1:1201] 1299.2 39.3 1833.1 2039.8 2727.6 ...
## ..$ age_category: Factor w/ 3 levels "young","old",..: 3 2 3 3 2 2 2 1 3 3
...
## $ Current or former < 1y:'data.frame': 593 obs. of 26 variables:
## ..$ id   : num [1:593] 3.00e+09 3.00e+09 4.83e+02 2.96e+03 2.94e+03 ...
## ..$ year  : Factor w/ 3 levels "1995","2000",..: 2 2 3 3 3 2 3 1 1 3 ...
## ..$ age   : int [1:593] 37 40 42 48 42 57 58 68 48 38 ...
## ..$ sex   : Factor w/ 2 levels "Male","Female": 1 2 2 2 1 1 2 1 2 1 ...
## ..$ smoker  : Factor w/ 3 levels "Never smoker",..: 2 2 2 2 2 2 2 2 2 2 ...
## ..$ sbp   : int [1:593] 132 108 99 105 128 146 139 132 114 118 ...
## ..$ dbp   : int [1:593] 82 70 60 73 84 84 84 72 72 74 ...
## ..$ histhtn  : Factor w/ 2 levels "Yes","No": 2 2 1 2 1 1 2 2 2 2 ...
## ..$ txhtn  : Factor w/ 2 levels "No","Yes": 1 1 1 1 1 2 1 1 1 1 ...
## ..$ chol  : num [1:593] 245 NA 116 162 218 209 231 198 192 169 ...
## ..$ hdl   : num [1:593] 59.8 68.9 49 63 27 ...
## ..$ triglyc  : num [1:593] 89 94 39 45 372 134 82 116 74 82 ...
## ..$ ldl   : num [1:593] 167.4 NA 59.2 90 NA ...
## ..$ histchol : Factor w/ 2 levels "Yes","No": 2 2 2 2 2 1 2 2 2 2 ...
## ..$ txchol  : Factor w/ 2 levels "No","Yes": 1 1 1 1 1 1 1 1 1 1 ...
```

```
## ..$ height  : num [1:593] 170 158 160 150 180 172 161 167 158 178 ...
## ..$ weight  : num [1:593] 70 43.5 66 60 88 75 76 80 50.4 67 ...
## ..$ bmi   : num [1:593] 24.2 17.4 25.8 26.7 27.2 ...
## ..$ phyact  : num [1:593] 553 387 274 426 391 ...
## ..$ pcs   : num [1:593] 43.4 57.3 54.8 50.3 51.9 ...
## ..$ mcs   : num [1:593] 62.6 47.9 49.6 45.6 57.3 ...
## ..$ cv   : Factor w/ 2 levels "No","Yes": 1 1 1 2 1 1 2 1 1 1 ...
## ..$ tocv  : num [1:593] 1906 3239 494 1367 3456 ...
## ..$ death  : Factor w/ 2 levels "No","Yes": 1 1 1 1 1 1 2 1 NA 2 ...
## ..$ todeath  : num [1:593] 858 878 1764 2443 2149 ...
## ..$ age_category: Factor w/ 3 levels "young","old",..: 1 1 2 2 2 2 2 3 2 1
...
## $ Former >= 1y   :'data.frame': 439 obs. of 26 variables:
## ..$ id   : num [1:439] 3.91e+03 1.00e+09 2.26e+03 3.95e+03 2.07e+03 ...
## ..$ year  : Factor w/ 3 levels "1995","2000",..: 3 1 3 3 3 1 3 2 1 3 ...
## ..$ age   : int [1:439] 43 70 54 68 70 56 52 70 56 50 ...
## ..$ sex   : Factor w/ 2 levels "Male","Female": 2 1 1 1 1 1 1 1 1 2 ...
## ..$ smoker  : Factor w/ 3 levels "Never smoker",..: 3 3 3 3 3 3 3 3 3 3 ...
## ..$ sbp   : int [1:439] 95 142 130 158 195 124 158 160 118 110 ...
## ..$ dbp   : int [1:439] 65 78 66 71 97 82 94 86 76 69 ...
## ..$ histhtn  : Factor w/ 2 levels "Yes","No": 2 2 2 1 1 2 1 2 2 2 ...
## ..$ txhtn  : Factor w/ 2 levels "No","Yes": 1 1 1 2 2 1 1 1 1 1 ...
## ..$ chol  : num [1:439] 194 188 268 209 182 254 251 230 187 173 ...
## ..$ hdl   : num [1:439] 75 35.8 37 52 34 ...
## ..$ triglyc  : num [1:439] 68 137 128 150 113 145 132 92 52 70 ...
## ..$ ldl   : num [1:439] 105 125 205 127 125 ...
## ..$ histchol : Factor w/ 2 levels "Yes","No": 2 2 2 2 2 1 1 1 2 2 ...
## ..$ txchol  : Factor w/ 2 levels "No","Yes": 1 1 1 1 1 1 1 2 1 1 ...
## ..$ height  : num [1:439] 160 160 163 156 156 155 174 170 169 160 ...
## ..$ weight  : num [1:439] 53 62 79 69 80 72 84 95.5 65 70 ...
## ..$ bmi   : num [1:439] 20.7 24.2 29.7 28.4 32.9 ...
```

```
## ..$ phyact  : num [1:439] 289 208 284.7 65.9 148.6 ...
## ..$ pcs   : num [1:439] 57.3 44 52 57.4 NA ...
## ..$ mcs   : num [1:439] 51.5 58.6 38.2 48.3 NA ...
## ..$ cv   : Factor w/ 2 levels "No","Yes": 1 1 1 1 1 1 1 2 1 1 ...
## ..$ tocv  : num [1:439] 2441 245 3055 941 774 ...
## ..$ death  : Factor w/ 2 levels "No","Yes": 1 1 1 1 1 1 1 1 1 1 ...
## ..$ todeath  : num [1:439] 336 2029 2856 1971 934 ...
## ..$ age_category: Factor w/ 3 levels "young","old",..: 2 3 2 3 3 2 2 3 2 2
...
# mean systolic blood pressure

mean(l$`Never smoker`$sbp)
## [1] NA
mean(l$`Never smoker`$sbp, na.rm = TRUE)
## [1] 132.0801
mean(l$`Current or former < 1y`$sbp)
## [1] NA
mean(l$`Current or former < 1y`$sbp, na.rm = TRUE)
## [1] 127.6081
mean(l$`Former >= 1y`$sbp)
## [1] 132.7882
# mean diastolic blood pressure

mean(l$`Never smoker`$dbp)
## [1] NA
mean(l$`Never smoker`$dbp, na.rm = TRUE)
## [1] 79.47412
mean(l$`Current or former < 1y`$dbp)
## [1] NA
mean(l$`Current or former < 1y`$dbp, na.rm = TRUE)
## [1] 78.77027
```

```
mean(l$`Former >= 1y`$dbp)
## [1] 81.15945
```

The smoker column has splitted the dataframe to 3 dataframes, one for never smoker with 1201 rows, one for Current or former < 1y with 593 rows, and the last one for Former >= 1y with 439 rows. These are the same number of rows produced by the table() function.

The mean systolic and diastolic blood pressure for Never smoker is 132.08 and 79.47, respectively. The mean systolic and diastolic blood pressure for Current or former < 1y is 127.61 and 78.77, respectively, while The mean systolic and diastolic blood pressure for Former >= 1y is 132.79 and 81.16, respectively.

## 6.5 QUANTILE() FUNCTION

The quantile() function produces sample quantiles corresponding to the given probabilities. The smallest observation (minimum) corresponds to a probability of 0 and the largest (maximum) to a probability of 1.

Example:

```
x<-1:100

x
## [1] 1 2 3 4 5 6 7 8 9 10 11 12 13 14 15 16 17 18
## [19] 19 20 21 22 23 24 25 26 27 28 29 30 31 32 33 34 35 36
## [37] 37 38 39 40 41 42 43 44 45 46 47 48 49 50 51 52 53 54
## [55] 55 56 57 58 59 60 61 62 63 64 65 66 67 68 69 70 71 72
## [73] 73 74 75 76 77 78 79 80 81 82 83 84 85 86 87 88 89 90
## [91] 91 92 93 94 95 96 97 98 99 100
quantile(x)
##  0% 25% 50% 75% 100%
## 1.00 25.75 50.50 75.25 100.00
summary(x)
## Min. 1st Qu. Median Mean 3rd Qu. Max.
```

```
## 1.00 25.75 50.50 50.50 75.25 100.00
```

The default quantile() function produces 0%, 25%, 50%, 75%, and 100%. 0% is minimum data value and 100% is maximum data value. 25% = 25.75 means that 25% of data values in x are lower than 25.75.

25% is 1st quantile, 50% = 2nd quantile = median and 75% = 3rd quantile. By using the summary(), the same values are reported.

Any quantile or percentile can be produced. For example, the 65% and 83% quantiles for x vector will be 65.35 and 83.17, respectively

```
x
## [1] 1 2 3 4 5 6 7 8 9 10 11 12 13 14 15 16 17 18
## [19] 19 20 21 22 23 24 25 26 27 28 29 30 31 32 33 34 35 36
## [37] 37 38 39 40 41 42 43 44 45 46 47 48 49 50 51 52 53 54
## [55] 55 56 57 58 59 60 61 62 63 64 65 66 67 68 69 70 71 72
## [73] 73 74 75 76 77 78 79 80 81 82 83 84 85 86 87 88 89 90
## [91] 91 92 93 94 95 96 97 98 99 100
quantile(x, probs = c(0.65,0.83))
## 65% 83%
## 65.35 83.17
```

### 6.5.1 Example From the Airquality Data

Find 45%, 66%, and 90% of ozone and temperature values.

```
str(airquality)
## 'data.frame': 153 obs. of 6 variables:
## $ Ozone : int 41 36 12 18 NA 28 23 19 8 NA ...
## $ Solar.R: int 190 118 149 313 NA NA 299 99 19 194 ...
## $ Wind : num 7.4 8 12.6 11.5 14.3 14.9 8.6 13.8 20.1 8.6 ...
## $ Temp : int 67 72 74 62 56 66 65 59 61 69 ...
## $ Month : int 5 5 5 5 5 5 5 5 5 5 ...
## $ Day : int 1 2 3 4 5 6 7 8 9 10 ...
```

**quantile**(airquality$Ozone, probs = **c**(0.45,0.66,0.90))

## Error in quantile.default(airquality$Ozone, probs = c(0.45, 0.66, 0.9)): missing values and NaN's not allowed if 'na.rm' is FALSE

**quantile**(airquality$Ozone, probs = **c**(0.45,0.66,0.90), na.rm = TRUE)

## 45% 66% 90%

## 28 45 87

**quantile**(airquality$Temp, probs = **c**(0.45,0.66,0.90))

## 45% 66% 90%

## 78 82 90

Because ozone column contains NA values, so we must add the argument na.em = TRUE to obtain the desired percentiles.

The 45%, 66%, and 90% for the ozone column is 28, 45, and 87, respectively. This means that 45% of ozone values in airquality data is less than 28, 66% of ozone values in airquality data is less than 45, and 90% of ozone values in airquality data is less than 87.

The 45%, 66%, and 90% for the temperature column is 78, 82, and 90, respectively. This means that 45% of temperature values in airquality data is less than 78, 66% of temperature values in airquality data is less than 82, and 90% of temperature values in airquality data is less than 90.

### 6.5.2 Example from the Regicor Data

Cut the regicor dataframe (2294 rows) into nearly 3 equal parts according to physical activity column (phyact). This means that the first third will contain the lowest 33% of physical activity values, the second third will contain the middle 33% of physical activity data and the last third will contain the highest 34% of physical activity data.

## *3 equal parts are 0–0.33,0.33–0.66,0.66–1*

**quantile**(regicor$phyact, probs = **c**(0,0.33,0.66,1))

## Error in quantile.default(regicor$phyact, probs = c(0, 0.33, 0.66, 1)): missing values and NaN's not allowed if 'na.rm' is FALSE

**quantile**(regicor$phyact, probs = **c**(0,0.33,0.66,1), na.rm = TRUE)

```
## 0% 33% 66% 100%
## 0.0000 206.8731 429.0508 5083.1868
regicor$phyact_cat<-cut(regicor$phyact, breaks = c(0,206.87,429.05,5083.19),

    include.lowest = TRUE, labels = c("third1","third2","third3"))

table(regicor$phyact_cat)
##
## third1 third2 third3
## 728 728 750
head(regicor)
##    id year age sex     smoker sbp dbp histhtn txhtn
## 6101  2265 2005 70 Female   Never smoker 138 75  No No
## 5762  1882 2005 56 Female   Never smoker 139 89  No No
## 2992 3000105616 2000 37 Male Current or former < 1y 132 82  No No
## 2611 3000103485 2000 69 Female   Never smoker 168 97  No No
## 2762 3000103963 2000 70 Female     <NA> NA NA  No No
## 1516 3000100883 2000 40 Female Current or former < 1y 108 70  No No
##  chol  hdl triglyc  ldl histchol txchol height weight  bmi
## 6101 294 57.00000  93 218.40000  No  No 160 64.0 25.00000
## 5762 220 50.00000  160 138.00000  No  No 163 67.0 25.21736
## 2992 245 59.80429  89 167.39571  No  No 170 70.0 24.22145
## 2611 168 53.17571  116 91.62429  No  No 147 68.0 31.46837
## 2762 NA  NA  NA  NA  <NA> <NA>  NA  NA  NA
## 1516 NA 68.90000  94  NA  No  No 158 43.5 17.42509
##  phyact pcs mcs cv  tocv death todeath age_category
## 6101 304.2000 54.455 58.918 No 1024.882 Yes 1299.16343  very old
## 5762 160.3000 58.165 47.995 No 2756.849 No 39.32629   old
## 2992 552.7912 43.429 62.585 No 1905.969 No 858.42203  young
## 2611 522.0000 54.325 57.900 No 1055.380 No 1833.07619  very old
```

```
## 2762 NA NA NA <NA> NA <NA> NA very old
## 1516 386.9505 57.315 47.869 No 3239.241 No 877.61155 young
## phyact_cat
## 6101 third2
## 5762 third1
## 2992 third3
## 2611 third3
## 2762 <NA>
## 1516 third2
summary(regicor$phyact)
## Min. 1st Qu. Median Mean 3rd Qu. Max. NA's
## 0.0 159.5 303.7 398.8 521.9 5083.2 88
```

The new created column "phyact_cat" contains728 rows for third1, 728 rows for third2, and 750 rows for third3. The head() function confirms the presence of this new column.

But these rows sum to 2206 rows and not 2294 rows of the original regicor data, because the original phyact column contains 88 NA values. Using the summary() function has confirmed that.

Using the split() function, we can now examine the properties of the first third to the last third

```
l<-split(regicor, regicor$phyact_cat)
```

```
str(l)
## List of 3
## $ third1:'data.frame': 728 obs. of 27 variables:
## ..$ id : num [1:728] 1.88e+03 3.95e+03 2.07e+03 1.00e+09 3.00e+09 ...
## ..$ year : Factor w/ 3 levels "1995","2000",..: 3 3 3 1 2 3 3 1 1 3 ...
## ..$ age : int [1:728] 56 68 70 70 57 52 62 68 56 60 ...
## ..$ sex : Factor w/ 2 levels "Male","Female": 2 1 1 1 1 1 2 1 1 2 ...
## ..$ smoker : Factor w/ 3 levels "Never smoker",..: 1 3 3 1 2 3 1 2 3 1 ...
```

```
## ..$ sbp   : int [1:728] 139 158 195 132 146 158 94 132 118 119 ...
## ..$ dbp   : int [1:728] 89 71 97 74 84 94 47 72 76 55 ...
## ..$ histhtn  : Factor w/ 2 levels "Yes","No": 2 1 1 2 1 1 2 2 2 2 ...
## ..$ txhtn  : Factor w/ 2 levels "No","Yes": 1 2 2 1 2 1 1 1 1 1 ...
## ..$ chol  : num [1:728] 220 209 182 183 209 251 335 198 187 244 ...
## ..$ hdl   : num [1:728] 50 52 34 34.7 42.2 ...
## ..$ triglyc  : num [1:728] 160 150 113 95 134 132 153 116 52 114 ...
## ..$ ldl   : num [1:728] 138 127 125 129 140 ...
## ..$ histchol : Factor w/ 2 levels "Yes","No": 2 2 2 2 1 1 1 2 2 2 ...
## ..$ txchol  : Factor w/ 2 levels "No","Yes": 1 1 1 1 1 1 2 1 1 1 ...
## ..$ height  : num [1:728] 163 156 156 162 172 174 153 167 169 158 ...
## ..$ weight  : num [1:728] 67 69 80 77 75 84 55 80 65 94 ...
## ..$ bmi   : num [1:728] 25.2 28.4 32.9 29.3 25.4 ...
## ..$ phyact  : num [1:728] 160.3 65.9 148.6 164 74.2 ...
## ..$ pcs   : num [1:728] 58.2 57.4 NA 57.2 56 ...
## ..$ mcs   : num [1:728] 48 48.3 NA 48.8 28.4 ...
## ..$ cv   : Factor w/ 2 levels "No","Yes": 1 1 1 1 1 1 1 1 1 1 ...
## ..$ tocv  : num [1:728] 2757 941 774 979 1364 ...
## ..$ death  : Factor w/ 2 levels "No","Yes": 1 1 1 1 1 1 1 1 1 1 ...
## ..$ todeath  : num [1:728] 39.3 1971.5 934.2 2692.4 1967.5 ...
## ..$ age_category: Factor w/ 3 levels "young","old",..: 2 3 3 3 2 2 3 3 2 2
...
## ..$ phyact_cat : Factor w/ 3 levels "third1","third2",..: 1 1 1 1 1 1 1 1 1
1 ...
## $ third2:'data.frame': 728 obs. of 27 variables:
## ..$ id  : num [1:728] 2.26e+03 3.00e+09 1.00e+09 3.91e+03 1.00e+09 ...
## ..$ year  : Factor w/ 3 levels "1995","2000",..: 3 2 1 3 1 3 3 3 3 3 ...
## ..$ age   : int [1:728] 70 40 53 43 70 54 42 54 48 48 ...
## ..$ sex   : Factor w/ 2 levels "Male","Female": 2 2 2 2 1 1 2 1 2 2 ...
## ..$ smoker  : Factor w/ 3 levels "Never smoker",..: 1 2 1 3 3 3 2 1 2 1 ...
## ..$ sbp   : int [1:728] 138 108 132 95 142 130 99 117 105 114 ...
## ..$ dbp   : int [1:728] 75 70 78 65 78 66 60 70 73 81 ...
```

```
## ..$ histhtn  : Factor w/ 2 levels "Yes","No": 2 2 2 2 2 2 1 2 2 1 ...
## ..$ txhtn  : Factor w/ 2 levels "No","Yes": 1 1 1 1 1 1 1 1 1 2 ...
## ..$ chol  : num [1:728] 294 NA 254 194 188 268 116 211 162 163 ...
## ..$ hdl   : num [1:728] 57 68.9 86 75 35.8 37 49 49 63 59 ...
## ..$ triglyc  : num [1:728] 93 94 NA 68 137 128 39 144 45 63 ...
## ..$ ldl   : num [1:728] 218 NA NA 105 125 ...
## ..$ histchol : Factor w/ 2 levels "Yes","No": 2 2 2 2 2 2 2 1 2 2 ...
## ..$ txchol  : Factor w/ 2 levels "No","Yes": 1 1 1 1 1 1 1 1 1 1 ...
## ..$ height  : num [1:728] 160 158 154 160 160 ...
## ..$ weight  : num [1:728] 64 43.5 45.8 53 62 79 66 65 60 69 ...
## ..$ bmi   : num [1:728] 25 17.4 19.2 20.7 24.2 ...
## ..$ phyact  : num [1:728] 304 387 324 289 208 ...
## ..$ pcs   : num [1:728] 54.5 57.3 49.5 57.3 44 ...
## ..$ mcs   : num [1:728] 58.9 47.9 57.4 51.5 58.6 ...
## ..$ cv   : Factor w/ 2 levels "No","Yes": 1 1 1 1 1 1 1 1 2 1 ...
## ..$ tocv  : num [1:728] 1025 3239 1974 2441 245 ...
## ..$ death  : Factor w/ 2 levels "No","Yes": 2 1 1 1 1 1 1 1 1 1 ...
## ..$ todeath  : num [1:728] 1299 878 2728 336 2029 ...
## ..$ age_category: Factor w/ 3 levels "young","old",..: 3 1 2 2 3 2 2 2 2 2
...
## ..$ phyact_cat : Factor w/ 3 levels "third1","third2",..: 2 2 2 2 2 2 2 2 2
2 ...
## $ third3:'data.frame': 750 obs. of 27 variables:
## ..$ id   : num [1:750] 3e+09 3e+09 3e+09 3e+09 3e+09 ...
## ..$ year  : Factor w/ 3 levels "1995","2000",..: 2 2 2 2 2 1 3 3 2 3 ...
## ..$ age   : int [1:750] 37 69 66 35 72 74 56 46 35 47 ...
## ..$ sex   : Factor w/ 2 levels "Male","Female": 1 2 1 2 2 2 1 2 2 1 ...
## ..$ smoker  : Factor w/ 3 levels "Never smoker",..: 2 1 1 1 1 1 NA 1 1 3 ...
## ..$ sbp   : int [1:750] 132 168 120 111 155 164 129 104 110 147 ...
## ..$ dbp   : int [1:750] 82 97 72 70 80 66 83 76 70 102 ...
## ..$ histhtn  : Factor w/ 2 levels "Yes","No": 2 2 1 2 1 2 2 1 2 1 ...
## ..$ txhtn  : Factor w/ 2 levels "No","Yes": 1 1 2 1 1 1 1 1 1 1 ...
```

```
## ..$ chol  : num [1:750] 245 168 298 238 NA 236 202 165 197 297 ...
## ..$ hdl   : num [1:750] 59.8 53.2 78.9 78.8 63.5 ...
## ..$ triglyc  : num [1:750] 89 116 71 108 106 77 85 48 73 112 ...
## ..$ ldl   : num [1:750] 167.4 91.6 204.9 137.6 NA ...
## ..$ histchol : Factor w/ 2 levels "Yes","No": 2 2 1 1 1 1 2 2 2 NA ...
## ..$ txchol  : Factor w/ 2 levels "No","Yes": 1 1 1 1 1 2 1 1 1 NA ...
## ..$ height  : num [1:750] 170 147 164 163 155 ...
## ..$ weight  : num [1:750] 70 68 79.2 76 69 65.5 64 58 84.5 68 ...
## ..$ bmi   : num [1:750] 24.2 31.5 29.4 28.6 28.7 ...
## ..$ phyact  : num [1:750] 553 522 478 793 556 ...
## ..$ pcs   : num [1:750] 43.4 54.3 59.6 51.5 NA ...
## ..$ mcs   : num [1:750] 62.6 57.9 40.8 51.1 NA ...
## ..$ cv   : Factor w/ 2 levels "No","Yes": 1 1 1 1 1 1 1 1 1 1 ...
## ..$ tocv  : num [1:750] 1906 1055.4 63.5 2788.6 413.9 ...
## ..$ death  : Factor w/ 2 levels "No","Yes": 1 1 2 1 1 1 1 1 NA 1 ...
## ..$ todeath  : num [1:750] 858 1833 2040 1763 3464 ...
## ..$ age_category: Factor w/ 3 levels "young","old",..: 1 3 3 1 3 3 2 2 1 2
...
## ..$ phyact_cat : Factor w/ 3 levels "third1","third2",..: 3 3 3 3 3 3 3 3 3
3 ...
# mean of total cholesterol

mean(l$third1$chol, na.rm = TRUE)
## [1] 215.148
mean(l$third3$chol, na.rm = TRUE)
## [1] 221.4134
# mean of weight

mean(l$third1$weight,na.rm = TRUE)
## [1] 74.8376
mean(l$third3$weight, na.rm = TRUE)
## [1] 72.71193
```

```
# percentage of smoking habits

prop.table(table(l$third1$smoker))*100
##
## Never smoker Current or former < 1y Former >= 1y
## 51.26050 30.39216 18.34734
prop.table(table(l$third3$smoker))*100
##
## Never smoker Current or former < 1y Former >= 1y
## 55.37415 25.98639 18.63946
# percentage of cardiovascular events

prop.table(table(l$third1$cv))*100
##
## No Yes
## 95.924309 4.075691
prop.table(table(l$third3$cv))*100
##
## No Yes
## 96.083916 3.916084
```

From the output, we see that the mean total cholesterol for third1 category (lowest physical activity group) is 215.148, while the mean total cholesterol for third3 category (highest physical activity group) is 221.4134.

We see that the mean weight for third1 category (lowest physical activity group) is 74.84, while the mean weight for third3 category (highest physical activity group) is 72.71.

51.26% of the third1 category are never smokers compared to 30.4% are current or former < 1y smokers and 18.34% are Former >= 1y smokers.

55.37% of the third3 category are never smokers compared to 25.99% are current or former < 1y smokers and 18.64% are Former >= 1y smokers.

# 6.6 MISSING VALUES

## 6.6.1 NA Values

NA = Not Available, is the first type of missing data in R. Any mathematical operation involving NA will return NA also. See the above examples when calculating the mean for columns containing NA values.

```
x<-c(1,2,NA,4)

x+3
## [1] 4 5 NA 7
x*3
## [1] 3 6 NA 12
x/3
## [1] 0.3333333 0.6666667  NA 1.3333333
mean(x)
## [1] NA
sd(x)
## [1] NA
median(x)
## [1] NA
```

All operations involving NA will yield NA values.

### *6.6.2. NaN Values*

NaN = Not a Number, is the second type of missing value in R. It can be created by certain mathematical operations that use 0 or Inf (positive infinity in R). Also, any mathematical operation involving NaN will produce NaN.

```
0/0
## [1] NaN
Inf/Inf
## [1] NaN
Inf-Inf
## [1] NaN
NaN*3
## [1] NaN
NaN/3
## [1] NaN
```

```
NaN+1
## [1] NaN
```

NaN value is also NA value but the converse is not true.

### 6.6.3 is.na() and is.nan() Functions

is.na() and is.nan() are two functions that test if the elements of a vector is NA or NaN, respectively. It will return a logical vector of the same length as the original vector but with TRUEs and FALSEs. As R treats TRUEs as ones and FALSEs as zeros so by summing that vector, you can know how many NAs or NaNs are present in your vector.
So this is a third method for checking NA values in your vector besides summary() and table() functions.
Simple examples are given below:

```
x<-c(1,2,3,NA,4,5,5,10,NA)

summary(x)
## Min. 1st Qu. Median Mean 3rd Qu. Max. NA’s
## 1.000 2.500 4.000 4.286 5.000 10.000 2
table(x, useNA = “ifany”)
## x
## 1 2 3 4 5 10 <NA>
## 1 1 1 1 2 1 2
is.na(x)
## [1] FALSE FALSE FALSE TRUE FALSE FALSE FALSE FALSE TRUE
sum(is.na(x))
## [1] 2
```

Using the three functions, summary, table, and sum(is.na) has confirmed that there are 2 NA values in x vector.
is.na() is not the same as is.nan()

```
x<-c(1,2,3,NA,4,5,5,10,NA, NaN, 11)

summary(x)
## Min. 1st Qu. Median Mean 3rd Qu. Max. NA’s
## 1.000 2.750 4.500 5.125 6.250 11.000 3
table(x, useNA = “ifany”)
```

```
## x
## 1 2 3 4 5 10 11 <NA> NaN
## 1 1 1 1 2 1 1 2 1
is.na(x)
## [1] FALSE FALSE FALSE TRUE FALSE FALSE FALSE FALSE TRUE TRUE FALSE
sum(is.na(x))
## [1] 3
is.nan(x)
## [1] FALSE FALSE FALSE FALSE FALSE FALSE FALSE FALSE FALSE TRUE FALSE
sum(is.nan(x))
## [1] 1
```

summary() function treats NA and NaN as NA. table() function differentiates between NA and NaN. sum(is.na(x)) counts NA and NaN collectively, while sum(is.nan(x)) counts NaN only.

Examples from the regicor data:

```
head(regicor)
##    id year age sex    smoker sbp dbp histhtn txhtn
## 6101  2265 2005 70 Female   Never smoker 138 75  No No
## 5762  1882 2005 56 Female   Never smoker 139 89  No No
## 2992 3000105616 2000 37 Male Current or former < 1y 132 82  No No
## 2611 3000103485 2000 69 Female   Never smoker 168 97  No No
## 2762 3000103963 2000 70 Female     <NA> NA NA  No No
## 1516 3000100883 2000 40 Female Current or former < 1y 108 70  No No
##  chol  hdl triglyc  ldl histchol txchol height weight  bmi
## 6101 294 57.00000  93 218.40000  No  No 160 64.0 25.00000
## 5762 220 50.00000  160 138.00000  No  No 163 67.0 25.21736
## 2992 245 59.80429  89 167.39571  No  No 170 70.0 24.22145
## 2611 168 53.17571  116 91.62429  No  No 147 68.0 31.46837
```

```
## 2762 NA NA NA NA <NA> <NA> NA NA NA
## 1516 NA 68.90000 94 NA No No 158 43.5 17.42509
## phyact pcs mcs cv tocv death todeath age_category
## 6101 304.2000 54.455 58.918 No 1024.882 Yes 1299.16343 very old
## 5762 160.3000 58.165 47.995 No 2756.849 No 39.32629 old
## 2992 552.7912 43.429 62.585 No 1905.969 No 858.42203 young
## 2611 522.0000 54.325 57.900 No 1055.380 No 1833.07619 very old
## 2762 NA NA NA <NA> NA <NA> NA very old
## 1516 386.9505 57.315 47.869 No 3239.241 No 877.61155 young
## phyact_cat
## 6101 third2
## 5762 third1
## 2992 third3
## 2611 third3
## 2762 <NA>
## 1516 third2
## smoker column

summary(regicor$smoker)
## Never smoker Current or former < 1y Former >= 1y
## 1201 593 439
## NA's
## 61
table(regicor$smoker, useNA = "ifany")
##
## Never smoker Current or former < 1y Former >= 1y
## 1201 593 439
## <NA>
## 61
sum(is.na(regicor$smoker))
## [1] 61
```

*## total cholesterol column*

```
summary(regicor$chol)
## Min. 1st Qu. Median Mean 3rd Qu. Max. NA's
## 95.0 189.0 215.0 218.8 245.0 488.0  101
table(regicor$chol, useNA = "ifany")
##
## 95 101 109 110 111 113 115 116 117 118 119 121 122
##  3  2  1  1  1  1  1  1  1  1  1  1  2
## 123 124 126 129 131 132 133 134 135 136 137 139 140
##  4  3  1  4  3  3  2  1  1  2  1  6  6
## 141 142 143 144 145 146 147 148 149 150 151 152 153
##  2  5  3  4  1  5  7 12  4  9  6  4  7
## 154 155 156 157 158 159 160 161 162 163 164 165 166
## 12  6 13  7  7  5 10  7  8  6  9 13 12
## 167 168 169 170 171 172 173 174 175 176 177 178 179
## 13 18 15 11 12  9 18 16 18 15 12 21  9
## 180 181 182 183 184 185 186 187 188 189 190 191 192
## 12 11 12 16 16  8 17 11 17 15 18 23 16
## 193 194 195 196 197 198 199 200 201 202 203 204 205
## 22 19 20 19 24 29 15 18 30 24 16 17 24
## 206 207 208 209 210 211 212 212.8 213 214 215 216 217
## 19 17 21 25 23 19 30  1 20 23 16 19 15
## 218 219 220 221 222 223 224 225 226 227 228 229 230
## 18 14 19 20 19 16 20 13 20 24 17 16 17
## 231 232 233 234 234.8 235 236 237 238 239 240 241 241.3
## 17 13 19 12  1 21 21 20 21 26 18 21  1
## 242 243 244 245 246 247 248 249 250 251 252 253 254
## 13 20 14 18 11 11 11 25 12 13 11  7 13
## 255 256 257 258 259 260 261 262 263 264 265 266 267
##  6 12  7 13 17 12 13 13 14  7 14 10  7
```

```
## 268 269 270 270.7 271 272 273 274 275 276 277 278 279
## 11 10 9 1 6 9 4 12 7 6 6 6 11
## 280 281 282 283 284 285 286 287 288 289 290 291 292
## 2 5 7 7 3 6 5 3 4 2 1 10 2
## 293 294 295 296 297 298 299 300 301 302 303 304 305
## 8 7 4 4 5 4 2 2 1 3 2 2 1
## 306 307 308 309 310 311 312 313 314 315 316 317 318
## 3 1 1 4 3 4 3 2 6 2 1 1 3
## 319 320 321 322 325 328 330 331 333 335 338 341 342
## 1 1 3 2 2 1 1 1 1 2 1 2 1
## 344 345 346 347 348 349 350 351 352 353 354 359 361
## 1 2 1 2 1 2 1 1 1 1 2 1 1
## 363 364 366 369 372 373 384 386 395 404 412 436 488
## 1 1 1 1 1 1 1 1 1 1 1 1 1
## <NA>
## 101
sum(is.na(regicor$chol))
## [1] 101
## to see total NA in regicor data

sum(is.na(regicor))
## [1] 1932
```

There are 61 NA values in smoker column, while there are 101 NA values in total cholesterol (chol) column.

The total NA values in regicor dataframe is 1932.

# CHAPTER 7

# SUBSETTING OBJECTS

## CONTENTS

There are three operators that can be used to extract subsets of R objects:

1. The [] operator always returns an object of the same class as the original. It can be used to select multiple elements of an object;
2. The [[]] operator is used to extract elements of a list or a data frame. It can only be used to extract a single element and the class of the returned object will not necessarily be a list or data frame; and
3. The $ operator is used to extract elements of a list or data frame by literal name. Its semantics are similar to that of [[.

Subsetting means extracting certain elements from a vector, list, dataframe, or matrix based on specific conditions

## 7.1 SUBSETTING VECTORS

The subsetting is performed by using indices which are 4 types:

1. Positive indices;
2. Negative indices;
3. Logical indices; and
4. Character indices.

Only use the [] operator to subset vectors

### 7.1.1 Positive Indices

The positive indices subset the vector according to its position

```
x<-1:10

x
## [1] 1 2 3 4 5 6 7 8 9 10
#extract first element

x[1]
## [1] 1
#extract second element
```

```
x[2]
## [1] 2
#extract third element

x[3]
## [1] 3
#extract third and fifth elements

x[c(3,5)]
## [1] 3 5
#extract third, fifth, and sixth elements

x[c(3,5,6)]
## [1] 3 5 6
#or

x[c(3,5:6)]
## [1] 3 5 6
#extract last four elements

x[c(7,8,9,10)]
## [1] 7 8 9 10
#or

x[7:10]
## [1] 7 8 9 10
#or

x[c(7:10)]
## [1] 7 8 9 10
```

To extract certain elements, just pass their position between square brackets. If you want to extract multiple items, use c() function or colon ":" operator.

### 7.1.2 Negative Indices

The negative indices subset the vector and exclude the element(s) at that position.

```
x<-1:10

x
## [1] 1 2 3 4 5 6 7 8 9 10
#extract all elements except the first element

x[-1]
## [1] 2 3 4 5 6 7 8 9 10
#extract all elements except the second element

x[-2]
## [1] 1 3 4 5 6 7 8 9 10
#extract all elements except the third element

x[-3]
## [1] 1 2 4 5 6 7 8 9 10
#extract all elements except the third and fifth elements

x[-c(3,5)]
## [1] 1 2 4 6 7 8 9 10
# or

x[c(-3,-5)]
## [1] 1 2 4 6 7 8 9 10
```

*#extract all elements except the third, fifth, and sixth elements*

```
x[-c(3,5,6)]
## [1] 1 2 4 7 8 9 10
#or

x[-c(3,5:6)]
## [1] 1 2 4 7 8 9 10
#extract all elements except the last four elements

x[-c(7,8,9,10)]
## [1] 1 2 3 4 5 6
#or

x[–7:–10]
## [1] 1 2 3 4 5 6
#or

x[c(–7:–10)]
## [1] 1 2 3 4 5 6
```

To extract certain elements using negative indices, pass the position of other elements, you do not want to extract, to negative indices between square brackets so they will be excluded. If you want to exclude multiple elements, use c() function or colon ":" operator.

### 7.1.3 Logical Indices

As said before in Chapter 4, commonly used logical operators:

- greater than ">"
- smaller than "<"
- greater than or equal ">="
- smaller than or equal "<="

- equal "=="
- nonequal "!="

Example of greater than ">" operator.

```
x<–1:100

x
## [1] 1 2 3 4 5 6 7 8 9 10 11 12 13 14 15 16 17 18
## [19] 19 20 21 22 23 24 25 26 27 28 29 30 31 32 33 34 35 36
## [37] 37 38 39 40 41 42 43 44 45 46 47 48 49 50 51 52 53 54
## [55] 55 56 57 58 59 60 61 62 63 64 65 66 67 68 69 70 71 72
## [73] 73 74 75 76 77 78 79 80 81 82 83 84 85 86 87 88 89 90
## [91] 91 92 93 94 95 96 97 98 99 100
y<-x[x>50]

y
## [1] 51 52 53 54 55 56 57 58 59 60 61 62 63 64 65 66 67 68 69
## [20] 70 71 72 73 74 75 76 77 78 79 80 81 82 83 84 85 86 87 88
## [39] 89 90 91 92 93 94 95 96 97 98 99 100
```

y is defined as values of x that are greater than 50.

Example of smaller than "<" operator

```
x<–1:100

x
## [1] 1 2 3 4 5 6 7 8 9 10 11 12 13 14 15 16 17 18
## [19] 19 20 21 22 23 24 25 26 27 28 29 30 31 32 33 34 35 36
## [37] 37 38 39 40 41 42 43 44 45 46 47 48 49 50 51 52 53 54
## [55] 55 56 57 58 59 60 61 62 63 64 65 66 67 68 69 70 71 72
## [73] 73 74 75 76 77 78 79 80 81 82 83 84 85 86 87 88 89 90
## [91] 91 92 93 94 95 96 97 98 99 100
```

```
y<-x[x<50]
```

```
y
## [1] 1 2 3 4 5 6 7 8 9 10 11 12 13 14 15 16 17 18 19 20 21 22 23 24 25
## [26] 26 27 28 29 30 31 32 33 34 35 36 37 38 39 40 41 42 43 44 45 46
47 48 49
```

y is defined as values of x that are smaller than 50.

Example of greater than or equal ">=" operator.

```
x<–1:100
```

```
x
## [1] 1 2 3 4 5 6 7 8 9 10 11 12 13 14 15 16 17 18
## [19] 19 20 21 22 23 24 25 26 27 28 29 30 31 32 33 34 35 36
## [37] 37 38 39 40 41 42 43 44 45 46 47 48 49 50 51 52 53 54
## [55] 55 56 57 58 59 60 61 62 63 64 65 66 67 68 69 70 71 72
## [73] 73 74 75 76 77 78 79 80 81 82 83 84 85 86 87 88 89 90
## [91] 91 92 93 94 95 96 97 98 99 100
y<-x[x >= 50]
```

```
y
## [1] 50 51 52 53 54 55 56 57 58 59 60 61 62 63 64 65 66 67 68
## [20] 69 70 71 72 73 74 75 76 77 78 79 80 81 82 83 84 85 86 87
## [39] 88 89 90 91 92 93 94 95 96 97 98 99 100
```

y is defined as values of x that are greater than or equal to 50.

Example of smaller than or equal "<=" operator.

```
x<–1:100
```

```
x
## [1] 1 2 3 4 5 6 7 8 9 10 11 12 13 14 15 16 17 18
## [19] 19 20 21 22 23 24 25 26 27 28 29 30 31 32 33 34 35 36
## [37] 37 38 39 40 41 42 43 44 45 46 47 48 49 50 51 52 53 54
## [55] 55 56 57 58 59 60 61 62 63 64 65 66 67 68 69 70 71 72
## [73] 73 74 75 76 77 78 79 80 81 82 83 84 85 86 87 88 89 90
## [91] 91 92 93 94 95 96 97 98 99 100
y<-x[x <= 50]

y
## [1] 1 2 3 4 5 6 7 8 9 10 11 12 13 14 15 16 17 18 19 20 21 22 23 24 25
## [26] 26 27 28 29 30 31 32 33 34 35 36 37 38 39 40 41 42 43 44 45 46 47
48 49 50
```

y is defined as values of x that are smaller than or equal to 50.

Example of equal "==" operator.

```
x<-1:100

x
## [1] 1 2 3 4 5 6 7 8 9 10 11 12 13 14 15 16 17 18
## [19] 19 20 21 22 23 24 25 26 27 28 29 30 31 32 33 34 35 36
## [37] 37 38 39 40 41 42 43 44 45 46 47 48 49 50 51 52 53 54
## [55] 55 56 57 58 59 60 61 62 63 64 65 66 67 68 69 70 71 72
## [73] 73 74 75 76 77 78 79 80 81 82 83 84 85 86 87 88 89 90
## [91] 91 92 93 94 95 96 97 98 99 100
y<-x[x == 50]

y
## [1] 50
```

y is defined as values of x that are equal to 50. y is a vector of 1 element.
Example of nonequal "!=" operator.

```
x<–1:100

x
## [1] 1 2 3 4 5 6 7 8 9 10 11 12 13 14 15 16 17 18
## [19] 19 20 21 22 23 24 25 26 27 28 29 30 31 32 33 34 35 36
## [37] 37 38 39 40 41 42 43 44 45 46 47 48 49 50 51 52 53 54
## [55] 55 56 57 58 59 60 61 62 63 64 65 66 67 68 69 70 71 72
## [73] 73 74 75 76 77 78 79 80 81 82 83 84 85 86 87 88 89 90
## [91] 91 92 93 94 95 96 97 98 99 100
y<-x[x != 50]

y
## [1] 1 2 3 4 5 6 7 8 9 10 11 12 13 14 15 16 17 18 19
## [20] 20 21 22 23 24 25 26 27 28 29 30 31 32 33 34 35 36 37 38
## [39] 39 40 41 42 43 44 45 46 47 48 49 51 52 53 54 55 56 57 58
## [58] 59 60 61 62 63 64 65 66 67 68 69 70 71 72 73 74 75 76 77
## [77] 78 79 80 81 82 83 84 85 86 87 88 89 90 91 92 93 94 95 96
## [96] 97 98 99 100
```

y is defined as all values of x except 50.
You can define many conditions with the and "&" operator.
Example of extracting the x elements that are in the range 50–80.

```
x<–1:100

x
## [1] 1 2 3 4 5 6 7 8 9 10 11 12 13 14 15 16 17 18
## [19] 19 20 21 22 23 24 25 26 27 28 29 30 31 32 33 34 35 36
```

```
## [37] 37 38 39 40 41 42 43 44 45 46 47 48 49 50 51 52 53 54
## [55] 55 56 57 58 59 60 61 62 63 64 65 66 67 68 69 70 71 72
## [73] 73 74 75 76 77 78 79 80 81 82 83 84 85 86 87 88 89 90
## [91] 91 92 93 94 95 96 97 98 99 100
y<-x[x >= 50 & x <= 80]

y
## [1] 50 51 52 53 54 55 56 57 58 59 60 61 62 63 64 65 66 67 68 69 70 71
72 73 74
## [26] 75 76 77 78 79 80
```

y is defined as all values of x that are within the range 50–80.

You can define non-continuous range with the or "|" operator.

Example of extracting the x elements that are in the range 1–50 and 80–100.

```
x<–1:100

x
## [1] 1 2 3 4 5 6 7 8 9 10 11 12 13 14 15 16 17 18
## [19] 19 20 21 22 23 24 25 26 27 28 29 30 31 32 33 34 35 36
## [37] 37 38 39 40 41 42 43 44 45 46 47 48 49 50 51 52 53 54
## [55] 55 56 57 58 59 60 61 62 63 64 65 66 67 68 69 70 71 72
## [73] 73 74 75 76 77 78 79 80 81 82 83 84 85 86 87 88 89 90
## [91] 91 92 93 94 95 96 97 98 99 100
y<-x[x <= 50 | x >= 80]

y
## [1] 1 2 3 4 5 6 7 8 9 10 11 12 13 14 15 16 17 18 19
## [20] 20 21 22 23 24 25 26 27 28 29 30 31 32 33 34 35 36 37 38
## [39] 39 40 41 42 43 44 45 46 47 48 49 50 80 81 82 83 84 85 86
## [58] 87 88 89 90 91 92 93 94 95 96 97 98 99 100
```

y is defined as all values of x that are within the range 1–50 and 80–100.

### 7.1.4 %in% Operator

To extract multiple elements, you can use the "%in%" operator.

Example of extracting the x elements that are equal to 10,20,30.

```
x<–1:100

x
## [1] 1 2 3 4 5 6 7 8 9 10 11 12 13 14 15 16 17 18
## [19] 19 20 21 22 23 24 25 26 27 28 29 30 31 32 33 34 35 36
## [37] 37 38 39 40 41 42 43 44 45 46 47 48 49 50 51 52 53 54
## [55] 55 56 57 58 59 60 61 62 63 64 65 66 67 68 69 70 71 72
## [73] 73 74 75 76 77 78 79 80 81 82 83 84 85 86 87 88 89 90
## [91] 91 92 93 94 95 96 97 98 99 100
y<-x[x %in% c(10,20,30)]

y
## [1] 10 20 30
```

Note that this operator will extract actual elements but the positive and negative indices extract the elements according to its position.

Example of difference between positive indices and %in% for actual values.

```
x<-seq(0,50,5)

x
## [1] 0 5 10 15 20 25 30 35 40 45 50
#extract the first, fifth, tenth elements

y<-x[c(1,5,10)]
```

```
y
## [1] 0 20 45
#extract numbers 1,5,10

y<-x[x %in% c(1,5,10)]

y
## [1] 5 10
```

Note that 1 is not an element of x so it is not returned by %in% operator.

### 7.1.5 Extracting with is.na() Function

Another useful functions is is.na() function for extracting NA values and !is.na() function for extracting non-missing values. There is no operator called (==NA).

A simple example

```
x<-c(1,2,5,6, NA, NA, 10,20, NA)

length(x)
## [1] 9
y<-x[is.na(x)]

y
## [1] NA NA NA
y<-x[!is.na(x)]

y
## [1] 1 2 5 6 10 20
```

There are 3 NA values in x.

### 7.1.6 Subsetting for NA Values in Airquality Dataframe.

It is the daily air quality measurements in New York, May to September 1973.

It is a dataframe with 153 observations (rows) on 6 variables (columns):

1. **Ozone:** Numeric, Mean ozone in parts per billion from 1300 to 1500 hours at Roosevelt Island
2. **Solar.R:** Numeric, Solar radiation in Langleys in the frequency band 4000–7700 Angstroms from 0800 to 1200 hours at Central Park
3. **Wind:** numeric, Average wind speed in miles per hour at 0700 and 1000 hours at LaGuardia Airport (LGA).
4. **Temp:** numeric, Maximum daily temperature in degrees Fahrenheit at La Guardia Airport.
5. **Month:** Numeric, Month (5–9)
6. **Day:** Numeric, Day of month (1–31)

*Reference:* Chambers, Cleveland, Kleiner, and Tukey (1983).

To define certain column of the dataframe, use the "$" operator.

```
library(datasets)

data("airquality")

str(airquality)
## 'data.frame': 153 obs. of 6 variables:
## $ Ozone : int 41 36 12 18 NA 28 23 19 8 NA ...
## $ Solar.R: int 190 118 149 313 NA NA 299 99 19 194 ...
## $ Wind : num 7.4 8 12.6 11.5 14.3 14.9 8.6 13.8 20.1 8.6 ...
## $ Temp : int 67 72 74 62 56 66 65 59 61 69 ...
## $ Month : int 5 5 5 5 5 5 5 5 5 5 ...
## $ Day : int 1 2 3 4 5 6 7 8 9 10 ...
# to extract NA values from the ozone column

x<-airquality$Ozone[is.na(airquality$Ozone)]
```

x

## [1] NA NA NA NA NA NA NA NA NA NA NA NA NA NA NA NA NA NA NA NA NA NA NA NA NA

## [26] NA NA NA NA NA NA NA NA NA NA NA NA

**length**(x)

## [1] 37

*# to extract non-missing values from the ozone column*

x<-airquality$Ozone[**!is.na**(airquality$Ozone)]

x

## [1] 41 36 12 18 28 23 19 8 7 16 11 14 18 14 34 6 30 11

## [19] 1 11 4 32 23 45 115 37 29 71 39 23 21 37 20 12 13 135

## [37] 49 32 64 40 77 97 97 85 10 27 7 48 35 61 79 63 16 80

## [55] 108 20 52 82 50 64 59 39 9 16 78 35 66 122 89 110 44 28

## [73] 65 22 59 23 31 44 21 9 45 168 73 76 118 84 85 96 78 73

## [91] 91 47 32 20 23 21 24 44 21 28 9 13 46 18 13 24 16 13

## [109] 23 36 7 14 30 14 18 20

**length**(x)

## [1] 116

We see that there are 37 NA values in the ozone column and 116 non-missing values. The total is 153 which is total number of rows in the dataframe.

Example of the airquality, Temp column

*# to extract NA values from the Temp column*

x<-airquality$Temp[**is.na**(airquality$Temp)]

x

## integer(0)

```
length(x)
## [1] 0
# to extract non-missing values from the Temp column

x<-airquality$Temp[!is.na(airquality$Temp)]

x
## [1] 67 72 74 62 56 66 65 59 61 69 74 69 66 68 58 64 66 57 68 62 59 73
61 61 57
## [26] 58 57 67 81 79 76 78 74 67 84 85 79 82 87 90 87 93 92 82 80 79 77
72 65 73
## [51] 76 77 76 76 76 75 78 73 80 77 83 84 85 81 84 83 83 88 92 92 89 82
73 81 91
## [76] 80 81 82 84 87 85 74 81 82 86 85 82 86 88 86 83 81 81 81 82 86 85
87 89 90
## [101] 90 92 86 86 82 80 79 77 79 76 78 78 77 72 75 79 81 86 88 97 94
96 94 91 92
## [126] 93 93 87 84 80 78 75 73 81 76 77 71 71 78 67 76 68 82 64 71 81
69 63 70 77
## [151] 75 76 68
length(x)
## [1] 153
```

We see that there are no NA (missing) values in the Temp column and all 153 values are actual numbers.

Example of the airquality, Solar.R column

```
# to extract NA values from the Solar.R column

x<-airquality$Solar.R[is.na(airquality$Solar.R)]

x
## [1] NA NA NA NA NA NA NA
```

```
length(x)
## [1] 7
# to extract non-missing values from the Solar.R column

x<-airquality$Solar.R[!is.na(airquality$Solar.R)]

x
## [1] 190 118 149 313 299 99 19 194 256 290 274 65 334 307 78 322 44 8
## [19] 320 25 92 66 266 13 252 223 279 286 287 242 186 220 264 127 273 291
## [37] 323 259 250 148 332 322 191 284 37 120 137 150 59 91 250 135 127 47
## [55] 98 31 138 269 248 236 101 175 314 276 267 272 175 139 264 175 291 48
## [73] 260 274 285 187 220 7 258 295 294 223 81 82 213 275 253 254 83 24
## [91] 77 255 229 207 222 137 192 273 157 64 71 51 115 244 190 259 36 255
## [109] 212 238 215 153 203 225 237 188 167 197 183 189 95 92 252 220 230 259
## [127] 236 259 238 24 112 237 224 27 238 201 238 14 139 49 20 193 145 191
## [145] 131 223
length(x)
## [1] 146
```

We see that there are 7 NA values in the Solar.R column and 146 non-missing values. The total is 153 which is total number of rows in the dataframe.

### 7.1.7 Character Indices

Character indices is used to extract certain elements from vectors if they have names.

```
x<-1:12
```

```
names(x)<-month.abb

x
## Jan Feb Mar Apr May Jun Jul Aug Sep Oct Nov Dec
## 1 2 3 4 5 6 7 8 9 10 11 12
y<-x["Mar"]

y
## Mar
## 3
y<-x[c("Apr","Jun","Oct")]

y
## Apr Jun Oct
## 4 6 10
```

x is a numeric vector with numbers from 1 to 12. Each number was given a month abbreviation name so we can extract certain numbers with their associated names.

If you want to subset a character vector, simply use the logical operators. Compare the following examples to the above examples.

```
x<-month.abb

x
## [1] "Jan" "Feb" "Mar" "Apr" "May" "Jun" "Jul" "Aug" "Sep" "Oct"
"Nov" "Dec"
y<-x[x == "Mar"]

y
## [1] "Mar"
```

```
y<-x[x %in% c("Apr","Jun","Oct")]

y
## [1] "Apr" "Jun" "Oct"
```

Example from the regicor dataframe discussed in Chapter 6.

We want to see the distribution of nonsmokers among both genders so we extract only the values of "Never smoker" from the smoker column.

```
library(compareGroups)
## Warning: package 'compareGroups' was built under R version 3.6.3
## Loading required package: SNPassoc
## Warning: package 'SNPassoc' was built under R version 3.6.3
## Loading required package: haplo.stats
## Warning: package 'haplo.stats' was built under R version 3.6.3
## Loading required package: survival
## Warning: package 'survival' was built under R version 3.6.3
## Loading required package: mvtnorm
## Loading required package: parallel
## Registered S3 method overwritten by 'SNPassoc':
## method   from
## summary.haplo.glm haplo.stats
data("regicor")

table(regicor$smoker)
##
##  Never smoker Current or former < 1y   Former >= 1y
##     1201     593     439
table(regicor$sex)
##
## Male Female
```

```
## 1101 1193
table(regicor$smoker == "Never smoker", regicor$sex)
##
## Male Female
## FALSE 770 262
## TRUE 301 900
prop.table(table(regicor$smoker == "Never smoker", regicor$sex),2)*100
##
## Male Female
## FALSE 71.89542 22.54733
## TRUE 28.10458 77.45267
```

For the last function, we used 2 to get proportions by column (sex) and multiplied by 100 to get percentages.

From the result, we see that 28% of males are nonsmokers and 72% are smokers. For females, 77.5% are nonsmokers and 22.5% are smokers.

## 7.2 SUBSETTING MATRICES

### 7.2.1 Numerical Indices

Matrices can be subsetted with numerical indices within square brackets like [1,2]. The first number indicates the row number and the second number is the column number.

Recall that matrices are filled column-wise.

```
x<-matrix(1:24, nrow = 3, ncol = 8)

x
## [,1] [,2] [,3] [,4] [,5] [,6] [,7] [,8]
## [1,] 1 4 7 10 13 16 19 22
## [2,] 2 5 8 11 14 17 20 23
## [3,] 3 6 9 12 15 18 21 24
```

*# to extract the number of first row and first column*

```
y<-x[1,1]

y
## [1] 1
```

*# to extract the number of first row and second column*

```
y<-x[1,2]

y
## [1] 4
```

*# to extract the number of second row and first column*

```
y<-x[2,1]

y
## [1] 2
```

*# to extract the number of third row and fifth column*

```
y<-x[3,5]

y
## [1] 15
```

To extract multiple elements use c() function

```
x<-matrix(1:24, nrow = 3, ncol = 8)

x
## [,1] [,2] [,3] [,4] [,5] [,6] [,7] [,8]
```

```
## [1,] 1 4 7 10 13 16 19 22
## [2,] 2 5 8 11 14 17 20 23
## [3,] 3 6 9 12 15 18 21 24
# to extract elements of the first and third rows and of the third and fifth columns

y<-x[c(1,3),c(3,5)]

y
##  [,1] [,2]
## [1,] 7 13
## [2,] 9 15
# to extract elements of the first row and of the third and fifth columns

y<-x[1,c(3,5)]

y
## [1] 7 13
```

To extract the whole row or whole column, leave the other number empty

```
x<-matrix(1:24, nrow = 3, ncol = 8)

x
##  [,1] [,2] [,3] [,4] [,5] [,6] [,7] [,8]
## [1,] 1 4 7 10 13 16 19 22
## [2,] 2 5 8 11 14 17 20 23
## [3,] 3 6 9 12 15 18 21 24
# to extract elements of the first row

y<-x[1,]
```

```
y
## [1] 1 4 7 10 13 16 19 22
# to extract elements of the fourth column

y<-x[,4]

y
## [1] 10 11 12
# to extract elements of the fourth and fifth columns, use the c() function

y<-x[,c(4,5)]

y
##  [,1] [,2]
## [1,] 10 13
## [2,] 11 14
## [3,] 12 15
```

### 7.2.2 Character Indices

You can extract matrices using character indices if they have names.

USPersonalExpenditure is a built-in numerical matrix in R. This data set consists of United States personal expenditures (in billions of dollars) in the categories; food and tobacco, household operation, medical, and health, personal care, and private education for the years 1940, 1945, 1950, 1955 and 1960. It has 5 rows and 5 columns.

```
library(datasets)

data("USPersonalExpenditure")
```

USPersonalExpenditure
## 1940 1945 1950 1955 1960
## Food and Tobacco 22.200 44.500 59.60 73.2 86.80
## Household Operation 10.500 15.500 29.00 36.5 46.20
## Medical and Health 3.530 5.760 9.71 14.0 21.10
## Personal Care 1.040 1.980 2.45 3.4 5.40
## Private Education 0.341 0.974 1.80 2.6 3.64
**dimnames**(USPersonalExpenditure)
## [[1]]
## [1] "Food and Tobacco" "Household Operation" "Medical and Health"
## [4] "Personal Care" "Private Education"
##
## [[2]]
## [1] "1940" "1945" "1950" "1955" "1960"
*# to extract food row*

x<-USPersonalExpenditure["Food and Tobacco",]

x
## 1940 1945 1950 1955 1960
## 22.2 44.5 59.6 73.2 86.8
*# to extract personal care row*

x<-USPersonalExpenditure["Personal Care",]

x
## 1940 1945 1950 1955 1960
## 1.04 1.98 2.45 3.40 5.40
*# to extract 1945 column*

```
x<-USPersonalExpenditure[,"1945"]

x
## Food and Tobacco Household Operation Medical and Health  Personal
Care
##   44.500   15.500   5.760   1.980
## Private Education
##   0.974
# to extract 1955 column

x<-USPersonalExpenditure[,"1955"]

x
## Food and Tobacco Household Operation Medical and Health  Personal
Care
##   73.2   36.5   14.0   3.4
## Private Education
##    2.6
```

Here we extracted certain columns or rows from the USPersonalExpenditure matrix. You will notice that the extracted column or row is now a vector and not a matrix. If you extracted multiple rows or columns, the result is a matrix.

To keep the extracted row or column in a matrix with 1 row or 1 column, use the argument, drop = FALSE. This will suppress the dropping of dimensions when extracting matrices.

```
USPersonalExpenditure
##      1940 1945 1950 1955 1960
## Food and Tobacco 22.200 44.500 59.60 73.2 86.80
## Household Operation 10.500 15.500 29.00 36.5 46.20
## Medical and Health 3.530 5.760 9.71 14.0 21.10
## Personal Care  1.040 1.980 2.45 3.4 5.40
```

## Private Education 0.341 0.974 1.80 2.6 3.64

*# to extract the value of first row and second column*

x<-USPersonalExpenditure[1,2]

x

## [1] 44.5

**class**(x)

## [1] "numeric"

*# to extract the value of first row and second column with drop = FALSE argument*

x<-USPersonalExpenditure[1,2, drop = FALSE]

x

## 1945

## Food and Tobacco 44.5

**class**(x)

## [1] "matrix"

*# to extract the values of the first row*

x<-USPersonalExpenditure[1,]

x

## 1940 1945 1950 1955 1960

## 22.2 44.5 59.6 73.2 86.8

**class**(x)

## [1] "numeric"

*# to extract the values of first row with drop = FALSE argument*

```
x<-USPersonalExpenditure[1,, drop = FALSE]

x
##     1940 1945 1950 1955 1960
## Food and Tobacco 22.2 44.5 59.6 73.2 86.8
class(x)
## [1] "matrix"
# to extract the values of 1955 column

x<-USPersonalExpenditure[,"1955"]

x
## Food and Tobacco Household Operation Medical and Health  Personal Care
##    73.2    36.5    14.0     3.4
## Private Education
##     2.6
class(x)
## [1] "numeric"
# to extract the values of 1955 column with drop = FALSE argument

x<-USPersonalExpenditure[,"1955", drop = FALSE]

x
##     1955
## Food and Tobacco 73.2
## Household Operation 36.5
## Medical and Health 14.0
## Personal Care  3.4
```

```
## Private Education 2.6
class(x)
## [1] "matrix"
```

Notice that even when extracting single values with drop = FALSE, a matrix will be created that consisted of 1 row and 1 column and 1 element only.

## 7.3 SUBSETTING LISTS

As we said in Chapter 4, lists are special type of vectors because:

1. Lists can contain objects of different classes;
2. Lists can group individual values as well as whole R vectors. The grouped vectors need not be of the same length.

Each of the three operators for subsetting, [],[[]], $ can be used with lists. Only $ is used for named lists.

### 7.3.1 Extracting with []

Example of a list with individual values

```
x<-list(10,"a",TRUE, FALSE, "b","c")

x
## [[1]]
## [1] 10
##
## [[2]]
## [1] "a"
##
## [[3]]
## [1] TRUE
##
## [[4]]
## [1] FALSE
```

```
##
## [[5]]
## [1] "b"
##
## [[6]]
## [1] "c"
```

*# to extract the first element as list*

```
y<-x[1]

y
## [[1]]
## [1] 10
class(y)
## [1] "list"
y<-x[[1]]

y
## [1] 10
class(y)
## [1] "numeric"
```

The list contains different elements, numeric, character, and logical.

Extracting the list with [] will give a smaller list while extracting the list with [[]] will give its elements (numeric, logical, character)

To extract multiple elements, use the c() function between square brackets [].

```
x<-list(10,"a",TRUE, FALSE, "b","c")
```

```
x
## [[1]]
## [1] 10
##
## [[2]]
## [1] "a"
##
## [[3]]
## [1] TRUE
##
## [[4]]
## [1] FALSE
##
## [[5]]
## [1] "b"
##
## [[6]]
## [1] "c"
# to extract multiple elements, use the c() function

y<-x[c(1,3)]

y
## [[1]]
## [1] 10
##
## [[2]]
## [1] TRUE
class(y)
## [1] "list"
y<-x[[c(1,3)]]
```

```
## Error in x[[c(1, 3)]]: subscript out of bounds
y<-x[[c(1,1)]]

y
## [1] 10
class(y)
## [1] "numeric"
y<-x[1][[1]]

y
## [1] 10
class(y)
## [1] "numeric"
```

### 7.3.2 Extracting with [[]]

To extract with c() function and double square brackets, it used to extract the certain nested elements of the list. For example, [[c(1,1)]] will extract the first element of the first element of the list, [[c(1,3)]] will extract the third element of the first element of the list and it gives an error in the above example because the first element of the list contains only one element.

### 7.3.3 Extracting with $

lists can have names for its elements. In this case, $ can also be used and it extracts the elements of the list as [[]]

```
x<-list(number = 10, character = "a", logical = TRUE)

x
## $number
## [1] 10
##
## $character
## [1] "a"
```

```
##
## $logical
## [1] TRUE
y<-x[1]

y
## $number
## [1] 10
class(y)
## [1] "list"
y<-x$character

y
## [1] "a"
class(y)
## [1] "character"
y<-x[[2]]

y
## [1] "a"
class(y)
## [1] "character"
```

Note that the result of $ operator is the same as [[]].

To extract multiple elements, we can use the c() function with positive indices or character indices

```
x<-list(number = 10, character = "a", logical = TRUE)

x
## $number
```

```
## [1] 10
##
## $character
## [1] "a"
##
## $logical
## [1] TRUE
y<-x[c(1,3)]

y
## $number
## [1] 10
##
## $logical
## [1] TRUE
class(y)
## [1] "list"
y<-x[c("number","logical")]

y
## $number
## [1] 10
##
## $logical
## [1] TRUE
class(y)
## [1] "list"
```

### 7.3.4 Extracting Lists with Whole Vectors

Example of a list with whole vectors

x<-**list**(state_names = state.name, numeric_vector = 1:100, letters = letters, month_names = month.name)

```
x
## $state_names
## [1] "Alabama" "Alaska" "Arizona" "Arkansas"
## [5] "California" "Colorado" "Connecticut" "Delaware"
## [9] "Florida" "Georgia" "Hawaii" "Idaho"
## [13] "Illinois" "Indiana" "Iowa" "Kansas"
## [17] "Kentucky" "Louisiana" "Maine" "Maryland"
## [21] "Massachusetts" "Michigan" "Minnesota" "Mississippi"
## [25] "Missouri" "Montana" "Nebraska" "Nevada"
## [29] "New Hampshire" "New Jersey" "New Mexico" "New York"
## [33] "North Carolina" "North Dakota" "Ohio" "Oklahoma"
## [37] "Oregon" "Pennsylvania" "Rhode Island" "South Carolina"
## [41] "South Dakota" "Tennessee" "Texas" "Utah"
## [45] "Vermont" "Virginia" "Washington" "West Virginia"
## [49] "Wisconsin" "Wyoming"
##
## $numeric_vector
## [1] 1 2 3 4 5 6 7 8 9 10 11 12 13 14 15 16 17 18
## [19] 19 20 21 22 23 24 25 26 27 28 29 30 31 32 33 34 35 36
## [37] 37 38 39 40 41 42 43 44 45 46 47 48 49 50 51 52 53 54
## [55] 55 56 57 58 59 60 61 62 63 64 65 66 67 68 69 70 71 72
## [73] 73 74 75 76 77 78 79 80 81 82 83 84 85 86 87 88 89 90
## [91] 91 92 93 94 95 96 97 98 99 100
##
## $letters
## [1] "a" "b" "c" "d" "e" "f" "g" "h" "i" "j" "k" "l" "m" "n" "o" "p" "q"
"r" "s"
## [20] "t" "u" "v" "w" "x" "y" "z"
##
## $month_names
```

```
## [1] "January" "February" "March" "April" "May" "June"
## [7] "July" "August" "September" "October" "November" "December"
```

The first element of the list is a character vector of length 50 that contains the 50 state names. The second element of the list is a numeric vector of length 100 that contains numbers from 1 to 100. The third element of the list is a character vector of small letters of the English alphabet. The fourth element is a character vector of the 12 month names.

use [] to extract certain elements of the list as smaller lists

```
x<-list(state_names = state.name, numeric_vector = 1:100, letters = letters,
month_names = month.name)
```

```
x
## $state_names
## [1] "Alabama" "Alaska" "Arizona" "Arkansas"
## [5] "California" "Colorado" "Connecticut" "Delaware"
## [9] "Florida" "Georgia" "Hawaii" "Idaho"
## [13] "Illinois" "Indiana" "Iowa" "Kansas"
## [17] "Kentucky" "Louisiana" "Maine" "Maryland"
## [21] "Massachusetts" "Michigan" "Minnesota" "Mississippi"
## [25] "Missouri" "Montana" "Nebraska" "Nevada"
## [29] "New Hampshire" "New Jersey" "New Mexico" "New York"
## [33] "North Carolina" "North Dakota" "Ohio" "Oklahoma"
## [37] "Oregon" "Pennsylvania" "Rhode Island" "South Carolina"
## [41] "South Dakota" "Tennessee" "Texas" "Utah"
## [45] "Vermont" "Virginia" "Washington" "West Virginia"
## [49] "Wisconsin" "Wyoming"
##
## $numeric_vector
## [1] 1 2 3 4 5 6 7 8 9 10 11 12 13 14 15 16 17 18
## [19] 19 20 21 22 23 24 25 26 27 28 29 30 31 32 33 34 35 36
```

```
## [37] 37 38 39 40 41 42 43 44 45 46 47 48 49 50 51 52 53 54
## [55] 55 56 57 58 59 60 61 62 63 64 65 66 67 68 69 70 71 72
## [73] 73 74 75 76 77 78 79 80 81 82 83 84 85 86 87 88 89 90
## [91] 91 92 93 94 95 96 97 98 99 100
##
## $letters
## [1] "a" "b" "c" "d" "e" "f" "g" "h" "i" "j" "k" "l" "m" "n" "o" "p" "q"
"r" "s"
## [20] "t" "u" "v" "w" "x" "y" "z"
##
## $month_names
## [1] "January" "February" "March"  "April"  "May"  "June"
## [7] "July"  "August" "September" "October" "November" "December"
y<-x[1]

y
## $state_names
## [1] "Alabama"  "Alaska"   "Arizona"  "Arkansas"
## [5] "California"  "Colorado"  "Connecticut" "Delaware"
## [9] "Florida"  "Georgia"  "Hawaii"   "Idaho"
## [13] "Illinois"  "Indiana"  "Iowa"   "Kansas"
## [17] "Kentucky"  "Louisiana"  "Maine"   "Maryland"
## [21] "Massachusetts" "Michigan"  "Minnesota"  "Mississippi"
## [25] "Missouri"  "Montana"  "Nebraska"  "Nevada"
## [29] "New Hampshire" "New Jersey"  "New Mexico"  "New York"
## [33] "North Carolina" "North Dakota" "Ohio"   "Oklahoma"
## [37] "Oregon"   "Pennsylvania" "Rhode Island" "South Carolina"
## [41] "South Dakota" "Tennessee"  "Texas"   "Utah"
## [45] "Vermont"  "Virginia"  "Washington"  "West Virginia"
## [49] "Wisconsin"  "Wyoming"
class(y)
```

```
## [1] "list"
# or using character indices

y<-x["state_names"]

y
## $state_names
## [1] "Alabama" "Alaska" "Arizona" "Arkansas"
## [5] "California" "Colorado" "Connecticut" "Delaware"
## [9] "Florida" "Georgia" "Hawaii" "Idaho"
## [13] "Illinois" "Indiana" "Iowa" "Kansas"
## [17] "Kentucky" "Louisiana" "Maine" "Maryland"
## [21] "Massachusetts" "Michigan" "Minnesota" "Mississippi"
## [25] "Missouri" "Montana" "Nebraska" "Nevada"
## [29] "New Hampshire" "New Jersey" "New Mexico" "New York"
## [33] "North Carolina" "North Dakota" "Ohio" "Oklahoma"
## [37] "Oregon" "Pennsylvania" "Rhode Island" "South Carolina"
## [41] "South Dakota" "Tennessee" "Texas" "Utah"
## [45] "Vermont" "Virginia" "Washington" "West Virginia"
## [49] "Wisconsin" "Wyoming"
class(y)
## [1] "list"
y<-x[c(1,4)]

y
## $state_names
## [1] "Alabama" "Alaska" "Arizona" "Arkansas"
## [5] "California" "Colorado" "Connecticut" "Delaware"
## [9] "Florida" "Georgia" "Hawaii" "Idaho"
## [13] "Illinois" "Indiana" "Iowa" "Kansas"
## [17] "Kentucky" "Louisiana" "Maine" "Maryland"
```

```
## [21] "Massachusetts" "Michigan"  "Minnesota"  "Mississippi"
## [25] "Missouri"  "Montana"  "Nebraska"  "Nevada"
## [29] "New Hampshire" "New Jersey"  "New Mexico"  "New York"
## [33] "North Carolina" "North Dakota" "Ohio"   "Oklahoma"
## [37] "Oregon"   "Pennsylvania" "Rhode Island" "South Carolina"
## [41] "South Dakota" "Tennessee"  "Texas"   "Utah"
## [45] "Vermont"  "Virginia"  "Washington"  "West Virginia"
## [49] "Wisconsin"  "Wyoming"
##
## $month_names
## [1] "January" "February" "March"  "April"  "May"  "June"
## [7] "July"  "August" "September" "October" "November" "December"
class(y)
## [1] "list"
# or using character indices

y<-x[c("state_names","month_names")]

y
## $state_names
## [1] "Alabama"  "Alaska"   "Arizona"  "Arkansas"
## [5] "California"  "Colorado"  "Connecticut" "Delaware"
## [9] "Florida"  "Georgia"  "Hawaii"   "Idaho"
## [13] "Illinois"  "Indiana"  "Iowa"   "Kansas"
## [17] "Kentucky"  "Louisiana"  "Maine"   "Maryland"
## [21] "Massachusetts" "Michigan"  "Minnesota"  "Mississippi"
## [25] "Missouri"  "Montana"  "Nebraska"  "Nevada"
## [29] "New Hampshire" "New Jersey"  "New Mexico"  "New York"
## [33] "North Carolina" "North Dakota" "Ohio"   "Oklahoma"
## [37] "Oregon"   "Pennsylvania" "Rhode Island" "South Carolina"
## [41] "South Dakota" "Tennessee"  "Texas"   "Utah"
```

```
## [45] "Vermont" "Virginia" "Washington" "West Virginia"
## [49] "Wisconsin" "Wyoming"
##
## $month_names
## [1] "January" "February" "March" "April" "May" "June"
## [7] "July" "August" "September" "October" "November" "December"
class(y)
## [1] "list"
```

Use $ to extract certain elements of the list as individual vectors (numeric, logical, character)

```
x<-list(state_names = state.name, numeric_vector = 1:100, letters = letters, month_names = month.name)

x
## $state_names
## [1] "Alabama" "Alaska" "Arizona" "Arkansas"
## [5] "California" "Colorado" "Connecticut" "Delaware"
## [9] "Florida" "Georgia" "Hawaii" "Idaho"
## [13] "Illinois" "Indiana" "Iowa" "Kansas"
## [17] "Kentucky" "Louisiana" "Maine" "Maryland"
## [21] "Massachusetts" "Michigan" "Minnesota" "Mississippi"
## [25] "Missouri" "Montana" "Nebraska" "Nevada"
## [29] "New Hampshire" "New Jersey" "New Mexico" "New York"
## [33] "North Carolina" "North Dakota" "Ohio" "Oklahoma"
## [37] "Oregon" "Pennsylvania" "Rhode Island" "South Carolina"
## [41] "South Dakota" "Tennessee" "Texas" "Utah"
## [45] "Vermont" "Virginia" "Washington" "West Virginia"
## [49] "Wisconsin" "Wyoming"
##
## $numeric_vector
```

```
## [1] 1 2 3 4 5 6 7 8 9 10 11 12 13 14 15 16 17 18
## [19] 19 20 21 22 23 24 25 26 27 28 29 30 31 32 33 34 35 36
## [37] 37 38 39 40 41 42 43 44 45 46 47 48 49 50 51 52 53 54
## [55] 55 56 57 58 59 60 61 62 63 64 65 66 67 68 69 70 71 72
## [73] 73 74 75 76 77 78 79 80 81 82 83 84 85 86 87 88 89 90
## [91] 91 92 93 94 95 96 97 98 99 100
##
## $letters
## [1] "a" "b" "c" "d" "e" "f" "g" "h" "i" "j" "k" "l" "m" "n" "o" "p" "q"
"r" "s"
## [20] "t" "u" "v" "w" "x" "y" "z"
##
## $month_names
## [1] "January" "February" "March"  "April"  "May"  "June"
## [7] "July"  "August" "September" "October" "November" "December"
y<-x$numeric_vector

y
## [1] 1 2 3 4 5 6 7 8 9 10 11 12 13 14 15 16 17 18
## [19] 19 20 21 22 23 24 25 26 27 28 29 30 31 32 33 34 35 36
## [37] 37 38 39 40 41 42 43 44 45 46 47 48 49 50 51 52 53 54
## [55] 55 56 57 58 59 60 61 62 63 64 65 66 67 68 69 70 71 72
## [73] 73 74 75 76 77 78 79 80 81 82 83 84 85 86 87 88 89 90
## [91] 91 92 93 94 95 96 97 98 99 100
class(y)
## [1] "integer"
y<-x$state_names

y
## [1] "Alabama"  "Alaska"  "Arizona"  "Arkansas"
## [5] "California"  "Colorado"  "Connecticut" "Delaware"
```

```
## [9] “Florida” “Georgia” “Hawaii” “Idaho”
## [13] “Illinois” “Indiana” “Iowa” “Kansas”
## [17] “Kentucky” “Louisiana” “Maine” “Maryland”
## [21] “Massachusetts” “Michigan” “Minnesota” “Mississippi”
## [25] “Missouri” “Montana” “Nebraska” “Nevada”
## [29] “New Hampshire” “New Jersey” “New Mexico” “New York”
## [33] “North Carolina” “North Dakota” “Ohio” “Oklahoma”
## [37] “Oregon” “Pennsylvania” “Rhode Island” “South Carolina”
## [41] “South Dakota” “Tennessee” “Texas” “Utah”
## [45] “Vermont” “Virginia” “Washington” “West Virginia”
## [49] “Wisconsin” “Wyoming”
class(y)
## [1] “character”
```

Use [[]] to extract certain elements of the list as individual vectors (numeric, logical, character)

```
x<-list(state_names = state.name, numeric_vector = 1:100, letters = letters,
month_names = month.name)
```

```
x
## $state_names
## [1] “Alabama” “Alaska” “Arizona” “Arkansas”
## [5] “California” “Colorado” “Connecticut” “Delaware”
## [9] “Florida” “Georgia” “Hawaii” “Idaho”
## [13] “Illinois” “Indiana” “Iowa” “Kansas”
## [17] “Kentucky” “Louisiana” “Maine” “Maryland”
## [21] “Massachusetts” “Michigan” “Minnesota” “Mississippi”
## [25] “Missouri” “Montana” “Nebraska” “Nevada”
## [29] “New Hampshire” “New Jersey” “New Mexico” “New York”
## [33] “North Carolina” “North Dakota” “Ohio” “Oklahoma”
## [37] “Oregon” “Pennsylvania” “Rhode Island” “South Carolina”
```

```
## [41] "South Dakota" "Tennessee" "Texas" "Utah"
## [45] "Vermont" "Virginia" "Washington" "West Virginia"
## [49] "Wisconsin" "Wyoming"
##
## $numeric_vector
## [1] 1 2 3 4 5 6 7 8 9 10 11 12 13 14 15 16 17 18
## [19] 19 20 21 22 23 24 25 26 27 28 29 30 31 32 33 34 35 36
## [37] 37 38 39 40 41 42 43 44 45 46 47 48 49 50 51 52 53 54
## [55] 55 56 57 58 59 60 61 62 63 64 65 66 67 68 69 70 71 72
## [73] 73 74 75 76 77 78 79 80 81 82 83 84 85 86 87 88 89 90
## [91] 91 92 93 94 95 96 97 98 99 100
##
## $letters
## [1] "a" "b" "c" "d" "e" "f" "g" "h" "i" "j" "k" "l" "m" "n" "o" "p" "q"
"r" "s"
## [20] "t" "u" "v" "w" "x" "y" "z"
##
## $month_names
## [1] "January" "February" "March" "April" "May" "June"
## [7] "July" "August" "September" "October" "November" "December"
# to extract the first element which is state_names, a character vector

y<-x[[1]]

y
## [1] "Alabama" "Alaska" "Arizona" "Arkansas"
## [5] "California" "Colorado" "Connecticut" "Delaware"
## [9] "Florida" "Georgia" "Hawaii" "Idaho"
## [13] "Illinois" "Indiana" "Iowa" "Kansas"
## [17] "Kentucky" "Louisiana" "Maine" "Maryland"
## [21] "Massachusetts" "Michigan" "Minnesota" "Mississippi"
```

```
## [25] "Missouri" "Montana" "Nebraska" "Nevada"
## [29] "New Hampshire" "New Jersey" "New Mexico" "New York"
## [33] "North Carolina" "North Dakota" "Ohio" "Oklahoma"
## [37] "Oregon" "Pennsylvania" "Rhode Island" "South Carolina"
## [41] "South Dakota" "Tennessee" "Texas" "Utah"
## [45] "Vermont" "Virginia" "Washington" "West Virginia"
## [49] "Wisconsin" "Wyoming"
class(y)
## [1] "character"
# or using character indices

y<-x[["state_names"]]

y
## [1] "Alabama" "Alaska" "Arizona" "Arkansas"
## [5] "California" "Colorado" "Connecticut" "Delaware"
## [9] "Florida" "Georgia" "Hawaii" "Idaho"
## [13] "Illinois" "Indiana" "Iowa" "Kansas"
## [17] "Kentucky" "Louisiana" "Maine" "Maryland"
## [21] "Massachusetts" "Michigan" "Minnesota" "Mississippi"
## [25] "Missouri" "Montana" "Nebraska" "Nevada"
## [29] "New Hampshire" "New Jersey" "New Mexico" "New York"
## [33] "North Carolina" "North Dakota" "Ohio" "Oklahoma"
## [37] "Oregon" "Pennsylvania" "Rhode Island" "South Carolina"
## [41] "South Dakota" "Tennessee" "Texas" "Utah"
## [45] "Vermont" "Virginia" "Washington" "West Virginia"
## [49] "Wisconsin" "Wyoming"
class(y)
## [1] "character"
# to extract the third element which is letters, a character vector
```

```
y<-x[[3]]

y
## [1] “a” “b” “c” “d” “e” “f” “g” “h” “i” “j” “k” “l” “m” “n” “o” “p” “q”
“r” “s”
## [20] “t” “u” “v” “w” “x” “y” “z”
class(y)
## [1] “character”
# or using character indices

y<-x[[“letters”]]

y
## [1] “a” “b” “c” “d” “e” “f” “g” “h” “i” “j” “k” “l” “m” “n” “o” “p” “q”
“r” “s”
## [20] “t” “u” “v” “w” “x” “y” “z”
class(y)
## [1] “character”
# use [[]] with c() function to extract nested elements

# to extract the third element of the second element (numeric_vector) which
is 3

y<-x[[c(2,3)]]

y
## [1] 3
class(y)
## [1] “integer”
# extract the third element of the fourth element (month_names) which is
“March”
```

```
y<-x[[c(4,3)]]

y
## [1] "March"
class(y)
## [1] "character"
```

When using [[]] with c() function to extract nested elements, you cannot use character indices.

## 7.4 SUBSETTING DATAFRAMES

As we said in Chapter 5, dataframes are special type of list where every element of the list or dataframe is a column. All columns of the dataframes have equal length and this length is the number of rows. The names of dataframe are column names which may be used for the dataframe extraction.

Example of dataframe extraction using airquality dataframe

### 7.4.1 Extraction Using[]

use [] to extract certain columns of the dataframe as smaller dataframes

```
library(datasets)

data("airquality")

str(airquality)
## 'data.frame': 153 obs. of 6 variables:
## $ Ozone : int 41 36 12 18 NA 28 23 19 8 NA ...
## $ Solar.R: int 190 118 149 313 NA NA 299 99 19 194 ...
## $ Wind : num 7.4 8 12.6 11.5 14.3 14.9 8.6 13.8 20.1 8.6 ...
## $ Temp : int 67 72 74 62 56 66 65 59 61 69 ...
```

```
## $ Month : int 5 5 5 5 5 5 5 5 5 5 ...
## $ Day : int 1 2 3 4 5 6 7 8 9 10 ...
names(airquality)
## [1] "Ozone" "Solar.R" "Wind" "Temp" "Month" "Day"
head(airquality)
## Ozone Solar.R Wind Temp Month Day
## 1 41  190 7.4 67  5 1
## 2 36  118 8.0 72  5 2
## 3 12  149 12.6 74  5 3
## 4 18  313 11.5 62  5 4
## 5 NA  NA 14.3 56  5 5
## 6 28  NA 14.9 66  5 6
# to extract the Ozone column

x<-airquality[1]

str(x)
## 'data.frame': 153 obs. of 1 variable:
## $ Ozone: int 41 36 12 18 NA 28 23 19 8 NA ...
class(x)
## [1] "data.frame"
# or using the character index

x<-airquality["Ozone"]

str(x)
## 'data.frame': 153 obs. of 1 variable:
## $ Ozone: int 41 36 12 18 NA 28 23 19 8 NA ...
class(x)
## [1] "data.frame"
# to extract the Month column
```

```
x<-airquality[5]

str(x)
## 'data.frame': 153 obs. of 1 variable:
## $ Month: int 5 5 5 5 5 5 5 5 5 5 ...
class(x)
## [1] "data.frame"
# or using the character index

x<-airquality["Month"]

str(x)
## 'data.frame': 153 obs. of 1 variable:
## $ Month: int 5 5 5 5 5 5 5 5 5 5 ...
class(x)
## [1] "data.frame"
# to extract multiple columns, Ozone and Wind, use the c() function

x<-airquality[c(1,3)]

str(x)
## 'data.frame': 153 obs. of 2 variables:
## $ Ozone: int 41 36 12 18 NA 28 23 19 8 NA ...
## $ Wind : num 7.4 8 12.6 11.5 14.3 14.9 8.6 13.8 20.1 8.6 ...
class(x)
## [1] "data.frame"
# or using the character index

x<-airquality[c("Ozone","Wind")]
```

**str**(x)

## 'data.frame': 153 obs. of 2 variables:

## $ Ozone: int 41 36 12 18 NA 28 23 19 8 NA ...

## $ Wind : num 7.4 8 12.6 11.5 14.3 14.9 8.6 13.8 20.1 8.6 ...

**class**(x)

## [1] "data.frame"

*# to extract the first row as a dataframe, use the row number with comma*

x<-airquality[1,]

**str**(x)

## 'data.frame': 1 obs. of 6 variables:

## $ Ozone : int 41

## $ Solar.R: int 190

## $ Wind : num 7.4

## $ Temp : int 67

## $ Month : int 5

## $ Day : int 1

**class**(x)

## [1] "data.frame"

*# to extract the first 10 rows as a dataframe, use the row number with comma and colon operator*

x<-airquality[1:10,]

**str**(x)

## 'data.frame': 10 obs. of 6 variables:

## $ Ozone : int 41 36 12 18 NA 28 23 19 8 NA

## $ Solar.R: int 190 118 149 313 NA NA 299 99 19 194

## $ Wind : num 7.4 8 12.6 11.5 14.3 14.9 8.6 13.8 20.1 8.6

## $ Temp : int 67 72 74 62 56 66 65 59 61 69

## $ Month : int 5 5 5 5 5 5 5 5 5 5

## $ Day : int 1 2 3 4 5 6 7 8 9 10

**class**(x)

## [1] "data.frame"

*# If you use that for columns, you will obtain vectors and not dataframes except when used for multiple # columns*

*# to extract the first column*

x<-airquality[,1]

**str**(x)

## int [1:153] 41 36 12 18 NA 28 23 19 8 NA ...

**class**(x)

## [1] "integer"

*# or use character indices*

x<-airquality[,'Ozone']

**str**(x)

## int [1:153] 41 36 12 18 NA 28 23 19 8 NA ...

**class**(x)

## [1] "integer"

*# When used for multiple columns, a smaller dataframe will be produced*

*# to extract the first 3 columns*

x<-airquality[,1:3]

**str**(x)

## 'data.frame': 153 obs. of 3 variables:

```
## $ Ozone : int 41 36 12 18 NA 28 23 19 8 NA ...
## $ Solar.R: int 190 118 149 313 NA NA 299 99 19 194 ...
## $ Wind : num 7.4 8 12.6 11.5 14.3 14.9 8.6 13.8 20.1 8.6 ...
class(x)
## [1] "data.frame"
# or use character indices

x<-airquality[,c("Ozone", "Solar.R", "Wind")]

str(x)
## 'data.frame': 153 obs. of 3 variables:
## $ Ozone : int 41 36 12 18 NA 28 23 19 8 NA ...
## $ Solar.R: int 190 118 149 313 NA NA 299 99 19 194 ...
## $ Wind : num 7.4 8 12.6 11.5 14.3 14.9 8.6 13.8 20.1 8.6 ...
class(x)
## [1] "data.frame"
# and you can use that to rearrange your columns

# to produce a smaller dataframe with Wind as first column, Ozone as second column, and Month as third # column

x<-airquality[,c("Wind", "Ozone", "Month")]

head(x)
## Wind Ozone Month
## 1 7.4 41  5
## 2 8.0 36  5
## 3 12.6 12  5
## 4 11.5 18  5
## 5 14.3 NA  5
## 6 14.9 28  5
```

```
# or use the column numbers

x<-airquality[,c(3,1,5)]

head(x)
## Wind Ozone Month
## 1 7.4 41  5
## 2 8.0 36  5
## 3 12.6 12  5
## 4 11.5 18  5
## 5 14.3 NA  5
## 6 14.9 28  5
```

### 7.4.2 Extraction Using $

use $ to extract certain columns of the dataframe as individual vectors (numeric, logical, character)

```
# to extract the Ozone column

x<-airquality$Ozone

str(x)
## int [1:153] 41 36 12 18 NA 28 23 19 8 NA ...
class(x)
## [1] "integer"
# to extract the Month column

x<-airquality$Month

str(x)
## int [1:153] 5 5 5 5 5 5 5 5 5 5 ...
class(x)
```

## [1] "integer"

### 7.4.3 Extraction Using [[]]

use [[]] to extract certain elements of the dataframe as individual vectors (numeric, logical, character)

```
# to extract the values of the first column or Ozone column

x<-airquality[[1]]

str(x)
## int [1:153] 41 36 12 18 NA 28 23 19 8 NA ...
class(x)
## [1] "integer"
# or using a character index

x<-airquality[["Ozone"]]

str(x)
## int [1:153] 41 36 12 18 NA 28 23 19 8 NA ...
class(x)
## [1] "integer"
# to extract the values of the fourth column or Temp column

x<-airquality[[4]]

str(x)
## int [1:153] 67 72 74 62 56 66 65 59 61 69 ...
class(x)
## [1] "integer"
# or using a character index
```

```
x<-airquality[["Temp"]]

str(x)
## int [1:153] 67 72 74 62 56 66 65 59 61 69 ...
class(x)
## [1] "integer"
# use [[]] with c() function to extract nested elements

# to extract the third value of the second column (Solar.R) which is 149

x<-airquality[[c(2,3)]]

x
## [1] 149
class(x)
## [1] "integer"
```

When using [[]] with c() function to extract nested elements, you cannot use character indices.

### 7.4.4 Using Logical Operators

Logical operators can be used to subset dataframes according to certain values.

Example using the airquality dataframe

```
summary(airquality)
## Ozone Solar.R Wind Temp
## Min. : 1.00 Min. : 7.0 Min. : 1.700 Min. :56.00
## 1st Qu.: 18.00 1st Qu.:115.8 1st Qu.: 7.400 1st Qu.:72.00
## Median : 31.50 Median :205.0 Median : 9.700 Median :79.00
## Mean : 42.13 Mean :185.9 Mean : 9.958 Mean :77.88
## 3rd Qu.: 63.25 3rd Qu.:258.8 3rd Qu.:11.500 3rd Qu.:85.00
```

```
## Max. :168.00 Max. :334.0 Max. :20.700 Max. :97.00
## NA's :37  NA's :7
##  Month   Day
## Min. :5.000 Min. : 1.0
## 1st Qu.:6.000 1st Qu.: 8.0
## Median :7.000 Median :16.0
## Mean :6.993 Mean :15.8
## 3rd Qu.:8.000 3rd Qu.:23.0
## Max. :9.000 Max. :31.0
##
```

*# we see that maximum Temp is 97, so to subset dataframe to get this row(s)*

```
x<-airquality[airquality$Temp==97,]

x
##  Ozone Solar.R Wind Temp Month Day
## 120 76  203 9.7 97  8 28
```

*# we see that third quartile of Ozone is 63.25, so to subset dataframe to get the rows that have Ozone larger than that value*

```
x<-airquality[airquality$Ozone > 63.25,]

x
##  Ozone Solar.R Wind Temp Month Day
## NA  NA  NA NA NA NA NA
## NA.1  NA  NA NA NA NA NA
## NA.2  NA  NA NA NA NA NA
## NA.3  NA  NA NA NA NA NA
## NA.4  NA  NA NA NA NA NA
## 30  115  223 5.7 79  5 30
## NA.5  NA  NA NA NA NA NA
```

```
## NA.6  NA  NA NA NA NA NA
## NA.7  NA  NA NA NA NA NA
## NA.8  NA  NA NA NA NA NA
## NA.9  NA  NA NA NA NA NA
## NA.10 NA  NA NA NA NA NA
## NA.11 NA  NA NA NA NA NA
## 40  71  291 13.8 90  6 9
## NA.12 NA  NA NA NA NA NA
## NA.13 NA  NA NA NA NA NA
## NA.14 NA  NA NA NA NA NA
## NA.15 NA  NA NA NA NA NA
## NA.16 NA  NA NA NA NA NA
## NA.17 NA  NA NA NA NA NA
## NA.18 NA  NA NA NA NA NA
## NA.19 NA  NA NA NA NA NA
## NA.20 NA  NA NA NA NA NA
## NA.21 NA  NA NA NA NA NA
## NA.22 NA  NA NA NA NA NA
## NA.23 NA  NA NA NA NA NA
## NA.24 NA  NA NA NA NA NA
## NA.25 NA  NA NA NA NA NA
## 62  135  269 4.1 84  7 1
## NA.26 NA  NA NA NA NA NA
## 66  64  175 4.6 83  7 5
## 68  77  276 5.1 88  7 7
## 69  97  267 6.3 92  7 8
## 70  97  272 5.7 92  7 9
## 71  85  175 7.4 89  7 10
## NA.27 NA  NA NA NA NA NA
## NA.28 NA  NA NA NA NA NA
## 80  79  187 5.1 87  7 19
```

```
## NA.29 NA  NA NA NA NA NA
## NA.30 NA  NA NA NA NA NA
## 85  80  294 8.6 86  7 24
## 86  108  223 8.0 85  7 25
## 89  82  213 7.4 88  7 28
## 91  64  253 7.4 83  7 30
## 96  78  NA 6.9 86  8 4
## 98  66  NA 4.6 87  8 6
## 99  122  255 4.0 89  8 7
## 100  89  229 10.3 90  8 8
## 101  110  207 8.0 90  8 9
## NA.31 NA  NA NA NA NA NA
## NA.32 NA  NA NA NA NA NA
## 106  65  157 9.7 80  8 14
## NA.33 NA  NA NA NA NA NA
## NA.34 NA  NA NA NA NA NA
## 117  168  238 3.4 81  8 25
## 118  73  215 8.0 86  8 26
## NA.35 NA  NA NA NA NA NA
## 120  76  203 9.7 97  8 28
## 121  118  225 2.3 94  8 29
## 122  84  237 6.3 96  8 30
## 123  85  188 6.3 94  8 31
## 124  96  167 6.9 91  9 1
## 125  78  197 5.1 92  9 2
## 126  73  183 2:8 93  9 3
## 127  91  189 4.6 93  9 4
## NA.36 NA  NA NA NA NA NA
# there are many NA values so to discard them, use !is.na() function

x<-airquality[airquality$Ozone > 63.25 & !is.na(airquality$Ozone),]
```

```
x
## Ozone Solar.R Wind Temp Month Day
## 30 115  223 5.7 79  5 30
## 40  71  291 13.8 90  6 9
## 62 135  269 4.1 84  7 1
## 66  64  175 4.6 83  7 5
## 68  77  276 5.1 88  7 7
## 69  97  267 6.3 92  7 8
## 70  97  272 5.7 92  7 9
## 71  85  175 7.4 89  7 10
## 80  79  187 5.1 87  7 19
## 85  80  294 8.6 86  7 24
## 86 108  223 8.0 85  7 25
## 89  82  213 7.4 88  7 28
## 91  64  253 7.4 83  7 30
## 96  78  NA 6.9 86  8 4
## 98  66  NA 4.6 87  8 6
## 99 122  255 4.0 89  8 7
## 100 89  229 10.3 90  8 8
## 101 110  207 8.0 90  8 9
## 106 65  157 9.7 80  8 14
## 117 168  238 3.4 81  8 25
## 118 73  215 8.0 86  8 26
## 120 76  203 9.7 97  8 28
## 121 118  225 2.3 94  8 29
## 122 84  237 6.3 96  8 30
## 123 85  188 6.3 94  8 31
## 124 96  167 6.9 91  9 1
## 125 78  197 5.1 92  9 2
## 126 73  183 2.8 93  9 3
```

```
## 127 91  189 4.6 93  9 4
# to subset the dataframe to get rows of days 30, 31

x<-airquality[airquality$Day %in% c(30,31),]

x
## Ozone Solar.R Wind Temp Month Day
## 30 115  223 5.7 79  5 30
## 31  37  279 7.4 76  5 31
## 61  NA  138 8.0 83  6 30
## 91  64  253 7.4 83  7 30
## 92  59  254 9.2 81  7 31
## 122 84  237 6.3 96  8 30
## 123 85  188 6.3 94  8 31
## 153 20  223 11.5 68  9 30
```

### 7.4.5 Other Complex Examples

To subset the dataframe to get rows of Ozone that are larger than 63.25 (third quartile of Ozone), temperature larger than 85 (third quartile of Temp), and Solar.R that are larger than 258.8 (third quartile of Solar.R). From the summary results, we see many NA values in Ozone and Solar.R columns so we filter for them too.

```
x<-airquality[airquality$Ozone > 63.25 & !is.na(airquality$Ozone) & airquality$Temp > 85 &

  airquality$Solar.R > 258.8 & !is.na(airquality$Solar.R),]

x
## Ozone Solar.R Wind Temp Month Day
## 40 71  291 13.8 90  6 9
## 68 77  276 5.1 88  7 7
## 69 97  267 6.3 92  7 8
```

```
## 70 97  272 5.7 92  7 9
## 85 80  294 8.6 86  7 24
```

We note that the subsetted rows fulfill the 3 conditions

Another interesting example

state.x77 is a matrix with 50 rows and 8 columns giving the following statistics in the respective columns:

1. **Population:** This estimate as of July 1, 1975
2. **Income:** per capita income (1974)
3. **Illiteracy:** illiteracy (1970, percent of population)
4. **Life Exp:** life expectancy in years (1969–71)
5. **Murder:** Murder and non-negligent manslaughter rate per 100,000 population (1976)
6. **HS Grad:** Percent high-school graduates (1970)
7. **Frost:** Mean number of days with minimum temperature below freezing (1931–1960) in capital or large city
8. **Area:** Land area in square miles

state.name: character vector giving the full state names.

state.region: factor giving the region (Northeast, South, North Central, West) that each state belongs to.

Note that all data are arranged according to alphabetical order of the state names.

We combine these three objects in a single dataframe called states and subset it according to some interesting characteristics.

```
library(datasets)

data(state)

states<-data.frame(state.name, state.region, state.x77)

head(states)
##   state.name state.region Population Income Illiteracy Life.Exp Murder
```

```
## Alabama Alabama South 3615 3624 2.1 69.05 15.1
## Alaska Alaska West 365 6315 1.5 69.31 11.3
## Arizona Arizona West 2212 4530 1.8 70.55 7.8
## Arkansas Arkansas South 2110 3378 1.9 70.66 10.1
## California California West 21198 5114 1.1 71.71 10.3
## Colorado Colorado West 2541 4884 0.7 72.06 6.8
## HS.Grad Frost Area
## Alabama 41.3 20 50708
## Alaska 66.7 152 566432
## Arizona 58.1 15 113417
## Arkansas 39.9 65 51945
## California 62.6 20 156361
## Colorado 63.9 166 103766
summary(states)
## state.name state.region Population Income
## Alabama : 1 Northeast : 9 Min. : 365 Min. :3098
## Alaska : 1 South :16 1st Qu.: 1080 1st Qu.:3993
## Arizona : 1 North Central:12 Median : 2838 Median :4519
## Arkansas : 1 West :13 Mean : 4246 Mean :4436
## California: 1 3rd Qu.: 4968 3rd Qu.:4814
## Colorado : 1 Max. :21198 Max. :6315
## (Other) :44
## Illiteracy Life.Exp Murder HS.Grad
## Min. :0.500 Min. :67.96 Min. : 1.400 Min. :37.80
## 1st Qu.:0.625 1st Qu.:70.12 1st Qu.: 4.350 1st Qu.:48.05
## Median :0.950 Median :70.67 Median : 6.850 Median :53.25
## Mean :1.170 Mean :70.88 Mean : 7.378 Mean :53.11
## 3rd Qu.:1.575 3rd Qu.:71.89 3rd Qu.:10.675 3rd Qu.:59.15
## Max. :2.800 Max. :73.60 Max. :15.100 Max. :67.30
##
## Frost Area
```

```
## Min. : 0.00 Min. : 1049
## 1st Qu.: 66.25 1st Qu.: 36985
## Median :114.50 Median : 54277
## Mean :104.46 Mean : 70736
## 3rd Qu.:139.75 3rd Qu.: 81163
## Max. :188.00 Max. :566432
##
str(states)
## 'data.frame': 50 obs. of 10 variables:
## $ state.name : Factor w/ 50 levels "Alabama","Alaska",..: 1 2 3 4 5 6 7
8 9 10 ...
## $ state.region: Factor w/ 4 levels "Northeast","South",..: 2 4 4 2 4 4 1 2
2 2 ...
## $ Population : num 3615 365 2212 2110 21198 ...
## $ Income : num 3624 6315 4530 3378 5114 ...
## $ Illiteracy : num 2.1 1.5 1.8 1.9 1.1 0.7 1.1 0.9 1.3 2 ...
## $ Life.Exp : num 69 69.3 70.5 70.7 71.7 ...
## $ Murder : num 15.1 11.3 7.8 10.1 10.3 6.8 3.1 6.2 10.7 13.9 ...
## $ HS.Grad : num 41.3 66.7 58.1 39.9 62.6 63.9 56 54.6 52.6 40.6 ...
## $ Frost : num 20 152 15 65 20 166 139 103 11 60 ...
## $ Area : num 50708 566432 113417 51945 156361 ...
```

After combining these three objects, the first column is the state.name, the second column is the state. region and the remaining 8 columns are the columns of the state.x77 matrix. The row names are the state names also.

From the summary() function, there are no NA values.

Question: Which state has the maximum murder rate (15.10)?

```
x<-states[states$Murder == 15.10, ]

x
## state.name state.region Population Income Illiteracy Life.Exp Murder
```

```
## Alabama Alabama  South  3615 3624  2.1 69.05 15.1
##   HS.Grad Frost Area
## Alabama 41.3 20 50708
```

Answer is Alabama state.

Question: Which state has the maximum high school graduation percentage (67.30)?

```
x<-states[states$HS.Grad == 67.30, ]

x
##  state.name state.region Population Income Illiteracy Life.Exp Murder
## Utah  Utah   West  1203 4022  0.6  72.9 4.5
##  HS.Grad Frost Area
## Utah 67.3 137 82096
```

Answer is Utah state.

Question: Which state has income > 4814 (third quartile) and murder rate < 4.35 (first quartile)?

```
x<-states[states$Income > 4814 & states$Murder < 4.35, ]

x
##    state.name state.region Population Income Illiteracy Life.Exp
## Connecticut Connecticut  Northeast  3100 5348  1.1 72.48
## North Dakota North Dakota North Central  637 5087  0.8 72.78
## Washington  Washington   West  3559 4864  0.6 71.72
##    Murder HS.Grad Frost Area
## Connecticut  3.1 56.0 139 4862
## North Dakota 1.4 50.3 186 69273
## Washington  4.3 63.5 32 66570
```

Answer is Connecticut, North Dakota, and Washington states.

Question: Which state has life expectancy > 71.89 (third quartile) and illiteracy > 1.575 (third quartile)?

```
x<-states[states$Life.Exp > 71.89 & states$Illiteracy > 1.575, ]

x
## state.name state.region Population Income Illiteracy Life.Exp Murder
## Hawaii Hawaii West 868 4963 1.9 73.6 6.2
## HS.Grad Frost Area
## Hawaii 61.9 0 6425
```

Answer is Hawaii state.

# 7.5 SORTING OBJECTS

## 7.5.1 Sort() Function

sort() function sorts any numeric vector from small to high. If the decreasing argument is set to TRUE, the numeric vector will be arranged from high to small.

```
x<-c(20,30,10,55,5)

x
## [1] 20 30 10 55 5
y<-sort(x)

y
## [1] 5 10 20 30 55
y<-sort(x, decreasing = TRUE)

y
```

```
## [1] 55 30 20 10 5
```

### 7.5.2 Order() Function

The order() function is used to give indices of sorted vector from small to high. If the decreasing argument is set to TRUE, the numeric vector will be arranged from high to small.

```
x<-c(20,30,10,55,5)

x
## [1] 20 30 10 55 5
y<-order(x)

y
## [1] 5 3 1 2 4
y<-order(x, decreasing = TRUE)

y
## [1] 4 2 1 3 5
```

The result of the first order() function states that element 5 of vector x (5) is smallest number and is put at position 1 followed by element 3 (10), element 1 (20), element 2 (30), then last one or largest number is element 4 (55).

We can use the order() function to subset and sort a vector in one line of code.

```
x<-c(20,30,10,55,5)

x
## [1] 20 30 10 55 5
x[order(x)]
## [1] 5 10 20 30 55
```

```
x[order(x, decreasing = TRUE)]
## [1] 55 30 20 10 5
```

We can use the above code to arrange columns or dataframes according to other columns.

Example, sort the state names according to murder rate.

```
head(states)
##   state.name state.region Population Income Illiteracy Life.Exp Murder
## Alabama  Alabama  South  3615 3624  2.1 69.05 15.1
## Alaska   Alaska   West  365 6315  1.5 69.31 11.3
## Arizona  Arizona   West  2212 4530  1.8 70.55 7.8
## Arkansas  Arkansas  South  2110 3378  1.9 70.66 10.1
## California California   West  21198 5114  1.1 71.71 10.3
## Colorado  Colorado   West  2541 4884  0.7 72.06 6.8
##   HS.Grad Frost Area
## Alabama  41.3 20 50708
## Alaska  66.7 152 566432
## Arizona  58.1 15 113417
## Arkansas  39.9 65 51945
## California 62.6 20 156361
## Colorado  63.9 166 103766
states$state.name[order(states$Murder)]
## [1] North Dakota South Dakota Iowa   Minnesota  Rhode Island
## [6] Maine   Nebraska  Wisconsin  Connecticut Massachusetts
## [11] New Hampshire Oregon   Washington  Kansas   Utah
## [16] Montana  New Jersey  Idaho   Vermont  Pennsylvania
## [21] Delaware  Hawaii   Oklahoma  West Virginia Colorado
## [26] Wyoming  Indiana  Ohio   Arizona  Maryland
## [31] Missouri  Virginia  New Mexico  Arkansas  California
## [36] Illinois  Kentucky  Florida  New York  Tennessee
```

## [41] Michigan North Carolina Alaska Nevada South Carolina

## [46] Texas Mississippi Louisiana Georgia Alabama

## 50 Levels: Alabama Alaska Arizona Arkansas California Colorado ... Wyoming

## *to order the whole dataframe, add comma after order() function to arrange its rows*

*# use head() function to see first 6 rows or smallest six states in murder rate*

**head**(states[**order**(states$Murder), ])

## state.name state.region Population Income Illiteracy Life.Exp

## North Dakota North Dakota North Central 637 5087 0.8 72.78

## South Dakota South Dakota North Central 681 4167 0.5 72.08

## Iowa Iowa North Central 2861 4628 0.5 72.56

## Minnesota Minnesota North Central 3921 4675 0.6 72.96

## Rhode Island Rhode Island Northeast 931 4558 1.3 71.90

## Maine Maine Northeast 1058 3694 0.7 70.39

## Murder HS.Grad Frost Area

## North Dakota 1.4 50.3 186 69273

## South Dakota 1.7 53.3 172 75955

## Iowa 2.3 59.0 140 55941

## Minnesota 2.3 57.6 160 79289

## Rhode Island 2.4 46.4 127 1049

## Maine 2.7 54.7 161 30920

*# use tail() function to see last 6 rows or largest six states in murder rate*

**tail**(states[**order**(states$Murder), ])

## state.name state.region Population Income Illiteracy

## South Carolina South Carolina South 2816 3635 2.3

## Texas Texas South 12237 4188 2.2

## Mississippi Mississippi South 2341 3098 2.4

## Louisiana Louisiana South 3806 3545 2.8

```
## Georgia    Georgia  South  4931 4091  2.0
## Alabama    Alabama  South  3615 3624  2.1
##     Life.Exp Murder HS.Grad Frost Area
## South Carolina 67.96 11.6 37.8 65 30225
## Texas    70.90 12.2 47.4 35 262134
## Mississippi  68.09 12.5 41.0 50 47296
## Louisiana   68.76 13.2 42.2 12 44930
## Georgia   68.54 13.9 40.6 60 58073
## Alabama   69.05 15.1 41.3 20 50708
```

The smallest state in murder rate is North Dakota and the largest state is Alabama.

Example, sort the state names according to life expectancy.

```
states$state.name[order(states$Life.Exp)]
## [1] South Carolina Mississippi Georgia  Louisiana  Nevada
## [6] Alabama  North Carolina Alaska   West Virginia Delaware
## [11] Virginia  Kentucky  Tennessee  Illinois  Maryland
## [16] Wyoming  New Mexico  Maine   Pennsylvania Arizona
## [21] New York  Montana  Michigan  Arkansas  Florida
## [26] Missouri  Ohio   Indiana  Texas   New Jersey
## [31] New Hampshire Oklahoma  Vermont  California  Washington
## [36] Massachusetts Idaho   Rhode Island Colorado  South Dakota
## [41] Oregon   Connecticut Wisconsin  Iowa   Kansas
## [46] Nebraska  North Dakota Utah   Minnesota  Hawaii
## 50 Levels: Alabama Alaska Arizona Arkansas California Colorado ... Wyoming
```

The smallest states in life expectancy is South Carolina and the largest state is Hawaii.

which.max() and which.min() functions can also give the index of the maximum and minimum numeric vectors.

For murder rates:

```
states$state.name[which.max(states$Murder)]
## [1] Alabama
## 50 Levels: Alabama Alaska Arizona Arkansas California Colorado ... Wyoming
states$state.name[which.min(states$Murder)]
## [1] North Dakota
## 50 Levels: Alabama Alaska Arizona Arkansas California Colorado ... Wyoming
```

The smallest state in murder rate is North Dakota and the largest state is Alabama.

For life expectancy:

```
states$state.name[which.max(states$Life.Exp)]
## [1] Hawaii
## 50 Levels: Alabama Alaska Arizona Arkansas California Colorado ... Wyoming
states$state.name[which.min(states$Life.Exp)]
## [1] South Carolina
## 50 Levels: Alabama Alaska Arizona Arkansas California Colorado ... Wyoming
```

The smallest states in life expectancy is South Carolina and the largest state is Hawaii.

### 7.5.3 Rank() Function

Another useful function is rank() function which returns the rank its argument. Here we used this function to create a rank column added to the states dataframe using $ argument.

```
head(states)
##   state.name state.region Population Income Illiteracy Life.Exp Murder
```

```
## Alabama  Alabama  South  3615 3624  2.1 69.05 15.1
## Alaska   Alaska   West  365 6315  1.5 69.31 11.3
## Arizona  Arizona   West  2212 4530  1.8 70.55 7.8
## Arkansas  Arkansas  South  2110 3378  1.9 70.66 10.1
## California California   West  21198 5114  1.1 71.71 10.3
## Colorado  Colorado   West  2541 4884  0.7 72.06 6.8
##   HS.Grad Frost Area
## Alabama  41.3 20 50708
## Alaska  66.7 152 566432
## Arizona  58.1 15 113417
## Arkansas  39.9 65 51945
## California 62.6 20 156361
## Colorado  63.9 166 103766
states$murder_rank<-rank(states$Murder)

head(states)
##   state.name state.region Population Income Illiteracy Life.Exp Murder
## Alabama  Alabama  South  3615 3624  2.1 69.05 15.1
## Alaska   Alaska   West  365 6315  1.5 69.31 11.3
## Arizona  Arizona   West  2212 4530  1.8 70.55 7.8
## Arkansas  Arkansas  South  2110 3378  1.9 70.66 10.1
## California California   West  21198 5114  1.1 71.71 10.3
## Colorado  Colorado   West  2541 4884  0.7 72.06 6.8
##   HS.Grad Frost Area murder_rank
## Alabama  41.3 20 50708  50.0
## Alaska  66.7 152 566432  43.0
## Arizona  58.1 15 113417  29.0
## Arkansas  39.9 65 51945  34.0
## California 62.6 20 156361  35.5
## Colorado  63.9 166 103766  25.0
tail(states)
```

```
##     state.name state.region Population Income Illiteracy Life.Exp
## Vermont    Vermont  Northeast  472 3907  0.6 71.64
## Virginia   Virginia   South  4981 4701  1.4 70.08
## Washington  Washington   West  3559 4864  0.6 71.72
## West Virginia West Virginia   South  1799 3617  1.4 69.48
## Wisconsin   Wisconsin North Central  4589 4468  0.7 72.48
## Wyoming    Wyoming   West  376 4566  0.6 70.29
##    Murder HS.Grad Frost Area murder_rank
## Vermont   5.5 57.1 168 9267   19
## Virginia   9.5 47.8 85 39780   32
## Washington  4.3 63.5 32 66570   13
## West Virginia 6.7 41.6 100 24070   24
## Wisconsin  3.0 54.5 149 54464   8
## Wyoming   6.9 62.9 173 97203   26
```

We see from the results that Alabama state has a murder_rank of 50.0 and because our dataframe has 50 rows so this state has the highest murder rate.

Note that there are some decimal ranks due to tied (repeated) values of murder rates.

### 7.5.4 Reorder() Function

Another useful function is the reorder() function which reorder the factor levels according to a calculated summary.

```
ordered_regions<-reorder(states$state.region,states$Murder,FUN = sum)

levels(ordered_regions)
## [1] "Northeast"  "North Central" "West"   "South"
l<-split(states, states$state.region)

sum(l$Northeast$Murder)
## [1] 42.5
sum(l$South$Murder)
```

```
## [1] 169.3
sum(l$`North Central`$Murder)
## [1] 63.3
sum(l$West$Murder)
## [1] 93.8
```

So Northeast has lowest sum of murder rates and South has the highest sum. We have confirmed the result using split() function.

Another example from the regicor dataframe.

```
library(compareGroups)

data("regicor")

# reorder the Recruitment year according to mean age of participants

ordered_year<-reorder(regicor$year, regicor$age, FUN = mean)

levels(ordered_year)
## [1] "1995" "2000" "2005"
```

1995 has lowest mean age, followed by 2000 and 2005 has largest mean age.

Reorder the smoking habits according to mean systolic blood pressure.

```
ordered_smoker<-reorder(regicor$smoker, regicor$sbp, FUN = mean)

levels(ordered_smoker)
## [1] "Former >= 1y"   "Never smoker"   "Current or former < 1y"
```

Former >= 1y has lowest mean systolic blood pressure and Current or former < 1y has the highest mean

reorder the smoking habits according to median LDL cholesterol

```
ordered_smoker<-reorder(regicor$smoker, regicor$ldl, FUN = median)

levels(ordered_smoker)
## [1] "Never smoker"   "Current or former < 1y" "Former >= 1y"
```

Never smoker has lowest median LDL cholesterol and Former >= 1y has the highest median.

## 7.6 REMOVING NA VALUES

As we saw in the above examples and in Chapter 6, is.na() function can be used to subset a vector for its missing values and its non-missing values.

Example:

```
x<-c(1,2,3,NA,4,5,5,10,NA)

x
## [1] 1 2 3 NA 4 5 5 10 NA
x[is.na(x)]
## [1] NA NA
x[!is.na(x)]
## [1] 1 2 3 4 5 5 10
```

Another function is complete.cases() that can be used to exclude NA values from a vector or dataframes.

```
x<-c(1,2,3,NA,4,5,5,10,NA)

x
## [1] 1 2 3 NA 4 5 5 10 NA
x[complete.cases(x)]
```

```
## [1] 1 2 3 4 5 5 10
x[!is.na(x)]
## [1] 1 2 3 4 5 5 10
```

We see that the result of subsetting using [!is.na(x)] is the same as using [complete.cases(x)].

When using complete.cases() for dataframes, use comma after the function to subset for rows containing no missing data.

```
library(datasets)

data("AirPassengers")

str(airquality)
## 'data.frame': 153 obs. of 6 variables:
## $ Ozone : int 41 36 12 18 NA 28 23 19 8 NA ...
## $ Solar.R: int 190 118 149 313 NA NA 299 99 19 194 ...
## $ Wind : num 7.4 8 12.6 11.5 14.3 14.9 8.6 13.8 20.1 8.6 ...
## $ Temp : int 67 72 74 62 56 66 65 59 61 69 ...
## $ Month : int 5 5 5 5 5 5 5 5 5 5 ...
## $ Day : int 1 2 3 4 5 6 7 8 9 10 ...
head(airquality)
## Ozone Solar.R Wind Temp Month Day
## 1 41  190 7.4 67  5 1
## 2 36  118 8.0 72  5 2
## 3 12  149 12.6 74  5 3
## 4 18  313 11.5 62  5 4
## 5 NA  NA 14.3 56  5 5
## 6 28  NA 14.9 66  5 6
# there are many missing values

x<-airquality[complete.cases(airquality),]
```

```
head(x)
## Ozone Solar.R Wind Temp Month Day
## 1 41  190 7.4 67  5 1
## 2 36  118 8.0 72  5 2
## 3 12  149 12.6 74  5 3
## 4 18  313 11.5 62  5 4
## 7 23  299 8.6 65  5 7
## 8 19  99 13.8 59  5 8
str(x)
## 'data.frame': 111 obs. of 6 variables:
## $ Ozone : int 41 36 12 18 23 19 8 16 11 14 ...
## $ Solar.R: int 190 118 149 313 299 99 19 256 290 274 ...
## $ Wind : num 7.4 8 12.6 11.5 8.6 13.8 20.1 9.7 9.2 10.9 ...
## $ Temp : int 67 72 74 62 65 59 61 69 66 68 ...
## $ Month : int 5 5 5 5 5 5 5 5 5 5 ...
## $ Day : int 1 2 3 4 7 8 9 12 13 14 ...
```

From the str() function, the airquality dataframe consists of 153 obs. (observations or rows). After excluding rows containing NA values, the resulting dataframe x consists of only 111 rows.

CHAPTER 8

# DATES AND TIMES

## CONTENTS

Dates and times in R are special objects that are used for time-series data or temporal data.

Dates are represented by the "Date" class and times are represented by the "POSIXct" or the "POSIXlt" class.

Dates are stored internally as the number of days since 1970-01-01.

Times are stored internally as the number of seconds since 1970-01-01 for "POSIXct" class or as the number of seconds in addition to other information for "POSIXlt" class.

## 8.1 DATES

Dates are represented by the Date class.

```
x<-Sys.Date()

x
## [1] "2020-04-26"
class(x)
## [1] "Date"
unclass(x)
## [1] 18378
```

As you can see, Sys.Date() returns the system date of the current date. It has a class "Date". Using unclass() function to see how the current date looks internally, it gives a large number (18377) which is the number of days since 1970-01-01.

### 8.1.1 as.Date() Function

The as.Date() function can be used to coerce character strings into "Date" class. It accepts the formats, Year-Month-Day or Year/Month/Day or it will give an error.

```
x<-as.Date("2018-03-21")

x
```

```
## [1] "2018-03-21"
class(x)
## [1] "Date"
x<-as.Date("2018/03/21")

x
## [1] "2018-03-21"
class(x)
## [1] "Date"
# 0 is not important for days and months

x<-as.Date("2018/3/21")

x
## [1] "2018-03-21"
class(x)
## [1] "Date"
x<-as.Date("2018/3/1")

x
## [1] "2018-03-01"
class(x)
## [1] "Date"
# errors with different formats

x<-as.Date("2018/21/03")
## Error in charToDate(x): character string is not in a standard unambiguous
format
x<-as.Date("2018/Mar/21")
## Error in charToDate(x): character string is not in a standard unambiguous
format
```

x<-**as.Date**("2018/March/21")

## Error in charToDate(x): character string is not in a standard unambiguous format

For dates before 1970-01-01, unclass() function will give negative numbers

x<-**as.Date**("1960-04-03")

x

## [1] "1960-04-03"

**class**(x)

## [1] "Date"

**unclass**(x)

## [1] –3560

x<-**as.Date**("1920-12-03")

x

## [1] "1920-12-03"

**class**(x)

## [1] "Date"

**unclass**(x)

## [1] –17926

From the function results, 1960-04-03 is before 1970-01-01 by 3560 days and 1920-12-03 is before 1970-01-01 by 17926 days.

**8.2 TIMES**

Times are stored in R with dates and are represented as class "POSIXct" or "POSIXlt" class.

x<-**Sys.time**()

```
x
## [1] "2020-04-26 17:01:17 EET"
class(x)
## [1] "POSIXct" "POSIXt"
unclass(x)
## [1] 1587913277
```

As you can see, Sys.time() returns the system date and time of the current date. It has a class "POSIXct". Using unclass() function to see how the current date looks internally, it gives a large number (1587840953) which is the number of seconds since 1970-01-01.

EET is the Eastern European Time zone. "POSIXt" is a common language between the two classes of the times with dates in R, "POSIXct" and "POSIXlt".

To see the "POSIXlt" class, use as.POSIXlt() function. The l instead of c in the class name indicates that it returns a list of components.

```
x<-Sys.time()


x
## [1] "2020-04-26 17:01:17 EET"
class(x)
## [1] "POSIXct" "POSIXt"
unclass(x)
## [1] 1587913278
y<-as.POSIXlt(x)

y
## [1] "2020-04-26 17:01:17 EET"
class(y)
## [1] "POSIXlt" "POSIXt"
```

**unclass**(y)

```
## $sec
## [1] 17.51696
##
## $min
## [1] 1
##
## $hour
## [1] 17
##
## $mday
## [1] 26
##
## $mon
## [1] 3
##
## $year
## [1] 120
##
## $wday
## [1] 0
##
## $yday
## [1] 116
##
## $isdst
## [1] 0
##
## $zone
## [1] "EET"
##
```

```
## $gmtoff
## [1] 7200
##
## attr(,"tzone")
## [1] ""  "EET" "EEST"
str(unclass(y))
## List of 11
## $ sec : num 17.5
## $ min : int 1
## $ hour : int 17
## $ mday : int 26
## $ mon : int 3
## $ year : int 120
## $ wday : int 0
## $ yday : int 116
## $ isdst : int 0
## $ zone : chr "EET"
## $ gmtoff: int 7200
## - attr(*, "tzone")= chr [1:3] "" "EET" "EEST"
```

Although y and x are similar in their values, y is a list of 11 elements, sec (second), min (minute), hour (hour), mday (monthday), mon(month = 3 as the first month January is assigned a 0 value), year,…

### 8.2.1 Difftime()

Another important function is difftime() that allows you to calculate the difference between two dates or dates-times in different units.

```
x<-as.Date(c("2018-03-21","2010-06-01"))

x
## [1] "2018-03-21" "2010-06-01"
```

```
class(x)
## [1] "Date"
## to see the difference between these two dates in days

difftime(x[1],x[2])
## Time difference of 2850 days
# In seconds

difftime(x[1],x[2], units = "secs")
## Time difference of 246240000 secs
# In minutes

difftime(x[1],x[2], units = "mins")
## Time difference of 4104000 mins
# In hours

difftime(x[1],x[2], units = "hours")
## Time difference of 68400 hours
# In weeks

difftime(x[1],x[2], units = "weeks")
## Time difference of 407.1429 weeks
```

## 8.3 LUBRIDATE PACKAGE

Lubridate package has a very useful function for creating dates or times with dates in R ("POSIXct" class). It has the general formula, ymd_hms, for year, month, day_hour, minute, second. By arranging the function as your strings, you can create date or date_time objects from any character value.

### 8.3.1 ymd() Function

Example, for format year-month-day, "1990-03-21".

```
library(lubridate)
##
## Attaching package: 'lubridate'
## The following object is masked from 'package:base':
##
## date
x<-ymd("1990-03-21")

x
## [1] "1990-03-21"
class(x)
## [1] "Date"
unclass(x)
## [1] 7384
```

Example, for format year-month-day, "1990-Mar-21".

```
library(lubridate)

x<-ymd("1990-Mar-21")

x
## [1] "1990-03-21"
class(x)
## [1] "Date"
unclass(x)
## [1] 7384
```

Example, for format year-month-day, "1990-March-21".

```
library(lubridate)
```

x<-**ymd**("1990-March-21")

x
## [1] "1990-03-21"
**class**(x)
## [1] "Date"
**unclass**(x)
## [1] 7384

Example, for format year-day-month, "1990-21-3".

**library**(lubridate)

x<-**ydm**("1990-21-3")

x
## [1] "1990-03-21"
**class**(x)
## [1] "Date"
**unclass**(x)
## [1] 7384

Example, for format month-year-day, "3-1990-21".

**library**(lubridate)

x<-**myd**("3-1990-21")

x
## [1] "1990-03-21"

```
class(x)
## [1] "Date"
unclass(x)
## [1] 7384
```

Example, for format month-day-year, "3-21-1990".

```
library(lubridate)

x<-mdy("3-21-1990")

x
## [1] "1990-03-21"
class(x)
## [1] "Date"
unclass(x)
## [1] 7384
```

Example, for format day-month-year, "21-3-1990".

```
library(lubridate)

x<-dmy("21-3-1990")

x
## [1] "1990-03-21"
class(x)
## [1] "Date"
unclass(x)
## [1] 7384
```

In all above examples, the unclass() function gives the same number, meaning that all these dates are the same date.

ymd() function also accepts different formats of the dates as well as numbers.

```
library(lubridate)

x<-ymd(c("2009-01-02", "2009.01.03","2009-1-4",

  "2009-1, 5", "200901-08", 20100121, 19910303))

x
## [1] "2009-01-02" "2009-01-03" "2009-01-04" "2009-01-05" "2009-01-08"
## [6] "2010-01-21" "1991-03-03"
class(x)
## [1] "Date"
as.POSIXlt(x)
## [1] "2009-01-02 UTC" "2009-01-03 UTC" "2009-01-04 UTC" "2009-01-05 UTC"
## [5] "2009-01-08 UTC" "2010-01-21 UTC" "1991-03-03 UTC"
```

Notice that the default time zone in ymd() function is UTC or Coordinated Universal Time. To add your time zone, add it to argument, tz.

```
x<-ymd(c("2009-01-02", "2009.01.03","2009-1-4",

  "2009-1, 5", "200901-08", 20100121, 19910303), tz= "EET")

x
## [1] "2009-01-02 EET" "2009-01-03 EET" "2009-01-04 EET" "2009-01-05 EET"
## [5] "2009-01-08 EET" "2010-01-21 EET" "1991-03-03 EET"
```

**class**(x)

## [1] "POSIXct" "POSIXt"

**as.POSIXlt**(x)

## [1] "2009-01-02 EET" "2009-01-03 EET" "2009-01-04 EET" "2009-01-05 EET"

## [5] "2009-01-08 EET" "2010-01-21 EET" "1991-03-03 EET"

## 8.3.2 ymd_hms() Function

For dates-times objects, arrange the function according to your data. It also accepts different formats:

x<-**ymd_hms**(**c**("2017-01-31 20:11:59","2010-04-01-12-00-00","2009-1-4 12-1-4",

"200901-08 1201-08","2009-1, 5 12:1, 5",20091010031025))

x

## [1] "2017-01-31 20:11:59 UTC" "2010-04-01 12:00:00 UTC"

## [3] "2009-01-04 12:01:04 UTC" "2009-01-08 12:01:08 UTC"

## [5] "2009-01-05 12:01:05 UTC" "2009-10-10 03:10:25 UTC"

**class**(x)

## [1] "POSIXct" "POSIXt"

x<-**ymd_hm**(**c**("2017-01-31 20:11","2010-04-01-12-00","2009-1-4 12-1",

"200901-08 1201","2009-1, 5 12:1",200910100310))

x

## [1] "2017-01-31 20:11:00 UTC" "2010-04-01 12:00:00 UTC"

## [3] "2009-01-04 12:01:00 UTC" "2009-01-08 12:01:00 UTC"

## [5] "2009-01-05 12:01:00 UTC" "2009-10-10 03:10:00 UTC"

**class**(x)

```
## [1] "POSIXct" "POSIXt"
x<-ymd_h(c("2017-01-31 20","2010-04-01-12","2009-1-4 12",

   "200901-08 12","2009-1, 5 12",2009101003))

x
## [1] "2017-01-31 20:00:00 UTC" "2010-04-01 12:00:00 UTC"
## [3] "2009-01-04 12:00:00 UTC" "2009-01-08 12:00:00 UTC"
## [5] "2009-01-05 12:00:00 UTC" "2009-10-10 03:00:00 UTC"
class(x)
## [1] "POSIXct" "POSIXt"
x<-mdy_h(c("01-31-2017 20","04-01-2010-12","1-4-2009 12",

   "01-08-2009 12","1, 5-2009 12",1010200903))

x
## [1] "2017-01-31 20:00:00 UTC" "2010-04-01 12:00:00 UTC"
## [3] "2009-01-04 12:00:00 UTC" "2009-01-08 12:00:00 UTC"
## [5] "2009-01-05 12:00:00 UTC" "2009-10-10 03:00:00 UTC"
class(x)
## [1] "POSIXct" "POSIXt"
```

### 8.3.3 day() and mday() Functions

day() to get the day component, mday() will get also the day component.

```
x
## [1] "2017-01-31 20:00:00 UTC" "2010-04-01 12:00:00 UTC"
## [3] "2009-01-04 12:00:00 UTC" "2009-01-08 12:00:00 UTC"
## [5] "2009-01-05 12:00:00 UTC" "2009-10-10 03:00:00 UTC"
day(x)
## [1] 31 1 4 8 5 10
mday(x)
```

## [1] 31 1 4 8 5 10

### 8.3.4 wday() Function

wday() will give the weekday as numbers. If the argument, label = TRUE, is added, the result will be abbreviated weekdays. If the argument, label = TRUE, abbr = FALSE, is added, the result will be full-named weekdays. The numbering of weekdays start at Sunday as number 1 and ends at Saturday as number 7.

```
x
## [1] "2017-01-31 20:00:00 UTC" "2010-04-01 12:00:00 UTC"
## [3] "2009-01-04 12:00:00 UTC" "2009-01-08 12:00:00 UTC"
## [5] "2009-01-05 12:00:00 UTC" "2009-10-10 03:00:00 UTC"
wday(x)
## [1] 3 5 1 5 2 7
wday(x,label = TRUE)
## [1] Tue Thu Sun Thu Mon Sat
## Levels: Sun < Mon < Tue < Wed < Thu < Fri < Sat
wday(x, label = TRUE, abbr = FALSE)
## [1] Tuesday Thursday Sunday Thursday Monday Saturday
## 7 Levels: Sunday < Monday < Tuesday < Wednesday < Thursday < ... <
Saturday
```

From the results, the first date is Tuesday, the second date is Thursday, the third date is Sunday, the fourth date is Thursday, the fifth date is Monday, and the sixth date is Saturday.

### 8.3.5 Month() Function

month() will give the month as numbers. If the argument, label = TRUE, is added, the result will be abbreviated months. If the argument, label = TRUE, abbr = FALSE, is added, the result will be full-named months. The numbering of months start at January as number 1 and ends at December as number 12.

```
x
## [1] “2017-01-31 20:00:00 UTC” “2010-04-01 12:00:00 UTC”
## [3] “2009-01-04 12:00:00 UTC” “2009-01-08 12:00:00 UTC”
## [5] “2009-01-05 12:00:00 UTC” “2009-10-10 03:00:00 UTC”
month(x)
## [1] 1 4 1 1 1 10
month(x, label = TRUE)
## [1] Jan Apr Jan Jan Jan Oct
## 12 Levels: Jan < Feb < Mar < Apr < May < Jun < Jul < Aug < Sep < ... < Dec
month(x, label = TRUE, abbr = FALSE)
## [1] January April January January January October
## 12 Levels: January < February < March < April < May < June < ... < December
```

From the results, the first date is January, the second date is April, the third date is January, the fourth date is January, the fifth date is January, and the sixth date is October.

### 8.3.6 Quarter() and Semester() Functions

quarter() function divides the year into fourths and semester divides the year into half. If the argument, with_year = TRUE, is added, the quarter or semester will be appended to its year.

```
x
## [1] “2017-01-31 20:00:00 UTC” “2010-04-01 12:00:00 UTC”
## [3] “2009-01-04 12:00:00 UTC” “2009-01-08 12:00:00 UTC”
## [5] “2009-01-05 12:00:00 UTC” “2009-10-10 03:00:00 UTC”
quarter(x)
## [1] 1 2 1 1 1 4
quarter(x, with_year = TRUE)
## [1] 2017.1 2010.2 2009.1 2009.1 2009.1 2009.4
semester(x)
```

```
## [1] 1 1 1 1 1 2
semester(x, with_year = TRUE)
## [1] 2017.1 2010.1 2009.1 2009.1 2009.1 2009.2
```

From the results, the first date is in first quarter, the second date is in second quarter, the third, fourth, and fifth dates are in the first quarter, and the sixth date is in the fourth quarter.

From the results, all the dates are in the first semester, and the sixth date is in the second semester.

### 8.3.7 Year() Function

year() function extracts the year component of a date

```
x
## [1] "2017-01-31 20:00:00 UTC" "2010-04-01 12:00:00 UTC"
## [3] "2009-01-04 12:00:00 UTC" "2009-01-08 12:00:00 UTC"
## [5] "2009-01-05 12:00:00 UTC" "2009-10-10 03:00:00 UTC"
year(x)
## [1] 2017 2010 2009 2009 2009 2009
```

## 8.4 MAKING DATES FROM INDIVIDUAL COMPONENTS

There are two important functions of lubridate:

4.1.make_date() function with its arguments, year, month, day.

4.2.make_datetime() function with its arguments, year, month, day, hour, min, sec, tz (time zone).

The first function make_date() will be used for the airquality dataframe and the second function make_datetime() will be used for the flights dataframe.

### 8.4.1 make_date() Function

The airquality dataframe contains daily air quality measurements in New York, May to September 1973. The dataframe contains only columns for month and day as evident from the str() function result.

```
library(datasets)

library(lubridate)

data("airquality")

str(airquality)
## 'data.frame': 153 obs. of 6 variables:
## $ Ozone : int 41 36 12 18 NA 28 23 19 8 NA ...
## $ Solar.R: int 190 118 149 313 NA NA 299 99 19 194 ...
## $ Wind : num 7.4 8 12.6 11.5 14.3 14.9 8.6 13.8 20.1 8.6 ...
## $ Temp : int 67 72 74 62 56 66 65 59 61 69 ...
## $ Month : int 5 5 5 5 5 5 5 5 5 5 ...
## $ Day : int 1 2 3 4 5 6 7 8 9 10 ...
# add date column

airquality$date<-make_date(year = 1973, month = airquality$Month,

    day = airquality$Day)

str(airquality)
## 'data.frame': 153 obs. of 7 variables:
## $ Ozone : int 41 36 12 18 NA 28 23 19 8 NA ...
## $ Solar.R: int 190 118 149 313 NA NA 299 99 19 194 ...
## $ Wind : num 7.4 8 12.6 11.5 14.3 14.9 8.6 13.8 20.1 8.6 ...
## $ Temp : int 67 72 74 62 56 66 65 59 61 69 ...
## $ Month : int 5 5 5 5 5 5 5 5 5 5 ...
## $ Day : int 1 2 3 4 5 6 7 8 9 10 ...
## $ date : Date, format: "1973-05-01" "1973-05-02" ...
head(airquality)
```

```
## Ozone Solar.R Wind Temp Month Day  date
## 1 41   190 7.4 67  5 1 1973-05-01
## 2 36   118 8.0 72  5 2 1973-05-02
## 3 12   149 12.6 74  5 3 1973-05-03
## 4 18   313 11.5 62  5 4 1973-05-04
## 5 NA   NA 14.3 56  5 5 1973-05-05
## 6 28   NA 14.9 66  5 6 1973-05-06
```

We now created another column called date. From this column, we can extract the weekday by wday() function, the quarter by quarter() function, and semester by semester() function.

```
head(airquality)
## Ozone Solar.R Wind Temp Month Day  date
## 1 41   190 7.4 67  5 1 1973-05-01
## 2 36   118 8.0 72  5 2 1973-05-02
## 3 12   149 12.6 74  5 3 1973-05-03
## 4 18   313 11.5 62  5 4 1973-05-04
## 5 NA   NA 14.3 56  5 5 1973-05-05
## 6 28   NA 14.9 66  5 6 1973-05-06
# add a column for weekday

airquality$weekday<-wday(airquality$date, label = TRUE)

table(airquality$weekday)
##
## Sun Mon Tue Wed Thu Fri Sat
## 22 21 22 22 22 22 22
# add a column for quarter

airquality$quarter<-quarter(airquality$date)
```

```
table(airquality$quarter)
##
## 2 3
## 61 92
# add a column for semester

airquality$semester<-semester(airquality$date)

table(airquality$semester)
##
## 1 2
## 61 92
# head() to see first 6 rows of the modified dataframe

# str() function to see data structure

head(airquality)
## Ozone Solar.R Wind Temp Month Day  date weekday quarter semester
## 1 41  190 7.4 67  5 1 1973-05-01  Tue  2  1
## 2 36  118 8.0 72  5 2 1973-05-02  Wed  2  1
## 3 12  149 12.6 74  5 3 1973-05-03  Thu  2  1
## 4 18  313 11.5 62  5 4 1973-05-04  Fri  2  1
## 5 NA  NA 14.3 56  5 5 1973-05-05  Sat  2  1
## 6 28  NA 14.9 66  5 6 1973-05-06  Sun  2  1
str(airquality)
## 'data.frame': 153 obs. of 10 variables:
## $ Ozone : int 41 36 12 18 NA 28 23 19 8 NA ...
## $ Solar.R : int 190 118 149 313 NA NA 299 99 19 194 ...
## $ Wind : num 7.4 8 12.6 11.5 14.3 14.9 8.6 13.8 20.1 8.6 ...
## $ Temp : int 67 72 74 62 56 66 65 59 61 69 ...
## $ Month : int 5 5 5 5 5 5 5 5 5 5 ...
```

```
## $ Day  : int 1 2 3 4 5 6 7 8 9 10 ...
## $ date : Date, format: "1973-05-01" "1973-05-02" ...
## $ weekday : Ord.factor w/ 7 levels "Sun"<"Mon"<"Tue"<..: 3 4 5 6 7 1
2 3 4 5 ...
## $ quarter : int 2 2 2 2 2 2 2 2 2 2 ...
## $ semester: int 1 1 1 1 1 1 1 1 1 1 ...
```

As we have modified this dataframe to include columns for weekdays, quarters, and semesters, we can now split it by semester and examine the properties of each semester.

Examine the median temperature for each semester.

```
# Exploring the data

head(airquality)
## Ozone Solar.R Wind Temp Month Day  date weekday quarter semester
## 1 41  190 7.4 67  5 1 1973-05-01  Tue  2  1
## 2 36  118 8.0 72  5 2 1973-05-02  Wed  2  1
## 3 12  149 12.6 74  5 3 1973-05-03  Thu  2  1
## 4 18  313 11.5 62  5 4 1973-05-04  Fri  2  1
## 5 NA  NA 14.3 56  5 5 1973-05-05  Sat  2  1
## 6 28  NA 14.9 66  5 6 1973-05-06  Sun  2  1
table(airquality$semester)
##
## 1 2
## 61 92
# splitting the data

x<-split(airquality,airquality$semester)

# Get the median temperature for each semester
```

```
# semester 1

median(x$`1`$Temp)
## [1] 73
# semester 2

median(x$`2`$Temp)
## [1] 82
```

The median temperature for semester 1 was 73, while for semester 2 was 82.

### 8.4.2 make_datetime() Function

New York City is serviced by three major airports:

1. John F. Kennedy International Airport (JFK)
2. LaGuardia Airport (LGA)
3. Newark International Airport (EWR)

flights dataframe from the nycflights13 package contains data for all flights that departed New York (i.e., JFK, LGA or EWR) in 2013.

It is a dataframe with 336,776 rows and 19 columns:

1. **year:** year of departure (2013)
2. **month:** month of departure
3. **day:** monthday of departure
4. **dep_time:** actual departure time (format HHMM or HMM), local tz
5. **sched_dep_time:** scheduled departure times (format HHMM or HMM), local tz
6. **dep_delay:** departure delays, in minutes. Negative times represent early departures/arrivals.
7. **arr_time:** actual arrival time (format HHMM or HMM), local tz
8. **sched_arr_time:** scheduled arrival times (format HHMM or HMM), local tz
9. **arr_delay:** arrival delays, in minutes. Negative times represent early departures/arrivals.

10. **carrier:** Two letter carrier abbreviation
11. **flight:** Flight number
12. **tailnum:** Plane tail number
13. **origin:** Origin
14. **dest:** destination
15. **air_time:** Amount of time spent in the air, in minutes
16. **distance:** Distance between airports, in miles.
17. **hour:** Time of scheduled departure broken into hour
18. **minute:** Time of scheduled departure broken into minutes
19. **time_hour:** Scheduled date and hour of the flight as a POSIXct date

New York has EST (Eastern Standard Time zone).

Add a column for departure date to flights dataframe.

```
library(nycflights13)
## Warning: package 'nycflights13' was built under R version 3.6.3
data("flights")


str(flights)
## Classes 'tbl_df', 'tbl' and 'data.frame': 336776 obs. of 19 variables:
## $ year   : int 2013 2013 2013 2013 2013 2013 2013 2013 2013 2013 ...
## $ month   : int 1 1 1 1 1 1 1 1 1 1 ...
## $ day   : int 1 1 1 1 1 1 1 1 1 1 ...
## $ dep_time  : int 517 533 542 544 554 554 555 557 557 558 ...
## $ sched_dep_time: int 515 529 540 545 600 558 600 600 600 600 ...
## $ dep_delay  : num 2 4 2 –1 –6 –4 –5 –3 –3 –2 ...
## $ arr_time  : int 830 850 923 1004 812 740 913 709 838 753 ...
## $ sched_arr_time: int 819 830 850 1022 837 728 854 723 846 745 ...
## $ arr_delay  : num 11 20 33 –18 –25 12 19 –14 –8 8 ...
## $ carrier  : chr "UA" "UA" "AA" "B6" ...
## $ flight  : int 1545 1714 1141 725 461 1696 507 5708 79 301 ...
## $ tailnum  : chr "N14228" "N24211" "N619AA" "N804JB" ...
## $ origin  : chr "EWR" "LGA" "JFK" "JFK" ...
```

```
## $ dest   : chr "IAH" "IAH" "MIA" "BQN" ...
## $ air_time  : num 227 227 160 183 116 150 158 53 140 138 ...
## $ distance  : num 1400 1416 1089 1576 762 ...
## $ hour   : num 5 5 5 5 6 5 6 6 6 6 ...
## $ minute  : num 15 29 40 45 0 58 0 0 0 0 ...
## $ time_hour  : POSIXct, format: "2013-01-01 05:00:00" "2013-01-01 05:00:00" ...
flights$dep_date<-make_datetime(year = flights$year, month = flights$month,

      day = flights$day, hour = flights$hour,

      min = flights$minute, tz = "EST")

str(flights)
## Classes 'tbl_df', 'tbl' and 'data.frame': 336776 obs. of 20 variables:
## $ year   : int 2013 2013 2013 2013 2013 2013 2013 2013 2013 2013 ...
## $ month   : int 1 1 1 1 1 1 1 1 1 1 ...
## $ day   : int 1 1 1 1 1 1 1 1 1 1 ...
## $ dep_time  : int 517 533 542 544 554 554 555 557 557 558 ...
## $ sched_dep_time: int 515 529 540 545 600 558 600 600 600 600 ...
## $ dep_delay  : num 2 4 2 –1 –6 –4 –5 –3 –3 –2 ...
## $ arr_time  : int 830 850 923 1004 812 740 913 709 838 753 ...
## $ sched_arr_time: int 819 830 850 1022 837 728 854 723 846 745 ...
## $ arr_delay  : num 11 20 33 –18 –25 12 19 –14 –8 8 ...
## $ carrier  : chr "UA" "UA" "AA" "B6" ...
## $ flight  : int 1545 1714 1141 725 461 1696 507 5708 79 301 ...
## $ tailnum  : chr "N14228" "N24211" "N619AA" "N804JB" ...
## $ origin  : chr "EWR" "LGA" "JFK" "JFK" ...
## $ dest   : chr "IAH" "IAH" "MIA" "BQN" ...
## $ air_time  : num 227 227 160 183 116 150 158 53 140 138 ...
## $ distance  : num 1400 1416 1089 1576 762 ...
```

```
## $ hour   : num 5 5 5 5 6 5 6 6 6 6 ...
## $ minute  : num 15 29 40 45 0 58 0 0 0 0 ...
## $ time_hour  : POSIXct, format: "2013-01-01 05:00:00" "2013-01-01 05:00:00" ...
## $ dep_date  : POSIXct, format: "2013-01-01 05:15:00" "2013-01-01 05:29:00" ...
head(flights)
## # A tibble: 6 x 20
## year month day dep_time sched_dep_time dep_delay arr_time sched_arr_time
## <int> <int> <int> <int>   <int>  <dbl> <int>   <int>
## 1 2013  1  1  517   515   2  830   819
## 2 2013  1  1  533   529   4  850   830
## 3 2013  1  1  542   540   2  923   850
## 4 2013  1  1  544   545  –1  1004   1022
## 5 2013  1  1  554   600  –6  812   837
## 6 2013  1  1  554   558  –4  740   728
## # … with 12 more variables: arr_delay <dbl>, carrier <chr>, flight <int>,
## # tailnum <chr>, origin <chr>, dest <chr>, air_time <dbl>, distance <dbl>,
## # hour <dbl>, minute <dbl>, time_hour <dttm>, dep_date <dttm>
```

We now created another column called dep_date. From this column, we can extract the weekday by wday() function, the quarter by quarter() function, and semester by semester() function.

```
head(flights)
## # A tibble: 6 x 20
## year month day dep_time sched_dep_time dep_delay arr_time sched_arr_time
## <int> <int> <int> <int>   <int>  <dbl> <int>   <int>
## 1 2013  1  1  517   515   2  830   819
## 2 2013  1  1  533   529   4  850   830
```

```
## 3 2013  1  1  542   540   2  923   850
## 4 2013  1  1  544   545  –1  1004  1022
## 5 2013  1  1  554   600  –6  812   837
## 6 2013  1  1  554   558  –4  740   728
## # … with 12 more variables: arr_delay <dbl>, carrier <chr>, flight <int>,
## # tailnum <chr>, origin <chr>, dest <chr>, air_time <dbl>, distance <dbl>,
## # hour <dbl>, minute <dbl>, time_hour <dttm>, dep_date <dttm>
# add a column for weekday

flights$weekday<-wday(flights$dep_date, label = TRUE)

table(flights$weekday)
##
## Sun Mon Tue Wed Thu Fri Sat
## 46357 50690 50422 50060 50219 50308 38720
# add a column for quarter

flights$quarter<-quarter(flights$dep_date)

table(flights$quarter)
##
##  1  2  3  4
## 80789 85369 86326 84292
# add a column for semester

flights$semester<-semester(flights$dep_date)

table(flights$semester)
##
##  1  2
```

```
## 166158 170618
# head() to see first 6 rows of the modified dataframe

# str() function to see data structure

head(flights)
## # A tibble: 6 x 23
## year month day dep_time sched_dep_time dep_delay arr_time sched_
arr_time
## <int> <int> <int> <int>   <int>  <dbl> <int>   <int>
## 1 2013  1  1  517   515   2  830   819
## 2 2013  1  1  533   529   4  850   830
## 3 2013  1  1  542   540   2  923   850
## 4 2013  1  1  544   545  –1  1004   1022
## 5 2013  1  1  554   600  –6  812   837
## 6 2013  1  1  554   558  –4  740   728
## # … with 15 more variables: arr_delay <dbl>, carrier <chr>, flight <int>,
## # tailnum <chr>, origin <chr>, dest <chr>, air_time <dbl>, distance
<dbl>,
## # hour <dbl>, minute <dbl>, time_hour <dttm>, dep_date <dttm>,
weekday <ord>,
## # quarter <int>, semester <int>
str(flights)
## Classes 'tbl_df', 'tbl' and 'data.frame': 336776 obs. of 23 variables:
## $ year   : int 2013 2013 2013 2013 2013 2013 2013 2013 2013 2013 ...
## $ month   : int 1 1 1 1 1 1 1 1 1 1 ...
## $ day   : int 1 1 1 1 1 1 1 1 1 1 ...
## $ dep_time  : int 517 533 542 544 554 554 555 557 557 558 ...
## $ sched_dep_time: int 515 529 540 545 600 558 600 600 600 600 ...
## $ dep_delay  : num 2 4 2 –1 –6 –4 –5 –3 –3 –2 ...
## $ arr_time  : int 830 850 923 1004 812 740 913 709 838 753 ...
## $ sched_arr_time: int 819 830 850 1022 837 728 854 723 846 745 ...
```

```
## $ arr_delay : num 11 20 33 –18 –25 12 19 –14 –8 8 ...
## $ carrier : chr "UA" "UA" "AA" "B6" ...
## $ flight : int 1545 1714 1141 725 461 1696 507 5708 79 301 ...
## $ tailnum : chr "N14228" "N24211" "N619AA" "N804JB" ...
## $ origin : chr "EWR" "LGA" "JFK" "JFK" ...
## $ dest : chr "IAH" "IAH" "MIA" "BQN" ...
## $ air_time : num 227 227 160 183 116 150 158 53 140 138 ...
## $ distance : num 1400 1416 1089 1576 762 ...
## $ hour : num 5 5 5 5 6 5 6 6 6 6 ...
## $ minute : num 15 29 40 45 0 58 0 0 0 0 ...
## $ time_hour : POSIXct, format: "2013-01-01 05:00:00" "2013-01-01 05:00:00" ...
## $ dep_date : POSIXct, format: "2013-01-01 05:15:00" "2013-01-01 05:29:00" ...
## $ weekday : Ord.factor w/ 7 levels "Sun"<"Mon"<"Tue"<..: 3 3 3 3 3 3 3 3 3 3 ...
## $ quarter : int 1 1 1 1 1 1 1 1 1 1 ...
## $ semester : int 1 1 1 1 1 1 1 1 1 1 ...
```

As we have modified this dataframe to include columns for weekdays, quarters, and semesters, we can now split it by weekdays and examine the properties of each weekday.

Examine the median arrival delay for each weekday.

```
# Exploring the data

head(flights)
## # A tibble: 6 x 23
## year month day dep_time sched_dep_time dep_delay arr_time sched_arr_time
## <int> <int> <int> <int>   <int>  <dbl> <int>   <int>
```

```
## 1 2013 1 1 517 515 2 830 819
## 2 2013 1 1 533 529 4 850 830
## 3 2013 1 1 542 540 2 923 850
## 4 2013 1 1 544 545 –1 1004 1022
## 5 2013 1 1 554 600 –6 812 837
## 6 2013 1 1 554 558 –4 740 728
## # … with 15 more variables: arr_delay <dbl>, carrier <chr>, flight <int>,
## # tailnum <chr>, origin <chr>, dest <chr>, air_time <dbl>, distance <dbl>,
## # hour <dbl>, minute <dbl>, time_hour <dttm>, dep_date <dttm>, weekday <ord>,
## # quarter <int>, semester <int>
table(flights$weekday)
##
## Sun Mon Tue Wed Thu Fri Sat
## 46357 50690 50422 50060 50219 50308 38720
# splitting the data

x<-split(flights, flights$weekday)

# Get the median arrival delay for each weekday

# Sunday and add na.rm = TRUE to remove NA values

median(x$Sun$arr_delay, na.rm = TRUE)
## [1] –6
# Monday

median(x$Mon$arr_delay, na.rm = TRUE)
## [1] –4
# Tuesday
```

```
median(x$Tue$arr_delay, na.rm = TRUE)
## [1] –5
# Wednesday

median(x$Wed$arr_delay, na.rm = TRUE)
## [1] –4
# Thursday

median(x$Thu$arr_delay, na.rm = TRUE)
## [1] –2
# Friday

median(x$Fri$arr_delay, na.rm = TRUE)
## [1] –3
# Saturday

median(x$Sat$arr_delay, na.rm = TRUE)
## [1] –9
```

The largest median arrival delay was for Thursday (–2) and the smallest median arrival delay was for Saturday (–9).

# CHAPTER 9

# IMPORTING DATA

## CONTENTS

# 9.1. IMPORTING COMMA SEPARATED VALUE FILES (.CSV EXTENSION) INTO R

## 9.1.1 read.csv() and read.csv2() Functions

If your values are separated with comma "," or semicolon ";", these files can be read with read.csv() and read.csv2() functions, respectively

read.csv() function is used to read files that their values are separated with ",", the decimal character is dot ".", and the first row is column names (header = TRUE) by its default behavior

read.csv2() function is used to read files that their values are separated with ";", the decimal character is comma ",", and the first row is column names (header = TRUE) by its default behavior

## 9.1.2 Example: Reading the Bank Data

Go to :http://archive.ics.uci.edu/ml/machine-learning-databases/00222/ and download bank.zip folder. This data is related to the marketing campaigns of a Portuguese bank for selling long-term deposits.

*Sources:* Moro, Cortez, and Rita, (2014).

From the bank.zip file, take bank file and put in your working folder. For example, I put it the computer E compartment and in a folder called test. To confirm that your bank file is in test folder, run the function, list.files("E:/test").

Open the file using WordPad or Notepad to see how the values are separated. You will see that the values are separated with semicolon ";" as the following image:

```
"age";"job";"marital";"education";"default";"balance";"housing
";"loan";"contact";"day";"month";"duration";"campaign";"pdays"
;"previous";"poutcome";"y"
30;"unemployed";"married";"primary";"no";1787;"no";"no";"cellu
lar";19;"oct";79;1;-1;0;"unknown";"no"
33;"services";"married";"secondary";"no";4789;"yes";"yes";"cel
lular";11;"may";220;1;339;4;"failure";"no"
35;"management";"single";"tertiary";"no";1350;"yes";"no";"cell
ular";16;"apr";185;1;330;1;"failure";"no"
30;"management";"married";"tertiary";"no";1476;"yes";"yes";"un
known";3;"jun";199;4;-1;0;"unknown";"no"
59;"blue-
collar";"married";"secondary";"no";0;"yes";"no";"unknown";5;"m
ay";226;1;-1;0;"unknown";"no"
35;"management";"single";"tertiary";"no";747;"no";"no";"cellul
ar";23;"feb";141;2;176;3;"failure";"no"
36;"self-
employed";"married";"tertiary";"no";307;"yes";"no";"cellular";
14;"may";341;1;330;2;"other";"no"
39;"technician";"married";"secondary";"no";147;"yes";"no";"cel
lular";6;"may";151;2;-1;0;"unknown";"no"
```

However, this file can be read with read.csv() or read.csv2() functions by manipulating its arguments.

```
# to locate your file

list.files("E:/test")
## [1] "bank.csv"     "bank2.csv"
## [3] "CTG (1).xls"    "CTG.xls"
## [5] "docword.nytimes.txt.gz"   "Online Retail.xlsx"
## [7] "Online.xlsx"     "readme.txt"
## [9] "Residential-Building-Data-Set.xlsx" "Unconfirmed 705073.crdownload"
## [11] "USCensus1990.data.txt"   "vocab.nytimes.txt"
## [13] "vocab.pubmed.txt"
# to read your file into R

dat<-read.csv("E:/test/bank.csv", sep = ";")

dat2<-read.csv2("E:/test/bank.csv", sep = ";", dec = ".")

identical(dat, dat2)
## [1] TRUE
```

The two functions have read the same file and produced the same dataset as evident from the identical function.

The produced dataset is a dataframe and so can be explored by different functions from the previous lessons

```
class(dat)
## [1] "data.frame"
str(dat)
```

```
## 'data.frame': 4521 obs. of 17 variables:
## $ age : int 30 33 35 30 59 35 36 39 41 43 ...
## $ job : Factor w/ 12 levels "admin.","blue-collar",..: 11 8 5 5 2 5 7 10 3 8 ...
## $ marital : Factor w/ 3 levels "divorced","married",..: 2 2 3 2 2 3 2 2 2 ...
## $ education: Factor w/ 4 levels "primary","secondary",..: 1 2 3 3 2 3 3 2 3 1 ...
## $ default : Factor w/ 2 levels "no","yes": 1 1 1 1 1 1 1 1 1 1 ...
## $ balance : int 1787 4789 1350 1476 0 747 307 147 221 –88 ...
## $ housing : Factor w/ 2 levels "no","yes": 1 2 2 2 2 1 2 2 2 2 ...
## $ loan : Factor w/ 2 levels "no","yes": 1 2 1 2 1 1 1 1 1 2 ...
## $ contact : Factor w/ 3 levels "cellular","telephone",..: 1 1 1 3 3 1 1 1 3 1 ...
## $ day : int 19 11 16 3 5 23 14 6 14 17 ...
## $ month : Factor w/ 12 levels "apr","aug","dec",..: 11 9 1 7 9 4 9 9 9 1 ...
## $ duration : int 79 220 185 199 226 141 341 151 57 313 ...
## $ campaign : int 1 1 1 4 1 2 1 2 2 1 ...
## $ pdays : int –1 339 330 –1 –1 176 330 –1 –1 147 ...
## $ previous : int 0 4 1 0 0 3 2 0 0 2 ...
## $ poutcome : Factor w/ 4 levels "failure","other",..: 4 1 1 4 4 1 2 4 4 1 ...
## $ y : Factor w/ 2 levels "no","yes": 1 1 1 1 1 1 1 1 1 1 ...
```

```
head(dat)
## age job marital education default balance housing loan contact day
## 1 30 unemployed married primary no 1787 no no cellular 19
## 2 33 services married secondary no 4789 yes yes cellular 11
## 3 35 management single tertiary no 1350 yes no cellular 16
## 4 30 management married tertiary no 1476 yes yes unknown 3
## 5 59 blue-collar married secondary no 0 yes no unknown 5
## 6 35 management single tertiary no 747 no no cellular 23
## month duration campaign pdays previous poutcome y
## 1 oct 79 1 –1 0 unknown no
```

```
## 2 may 220 1 339 4 failure no
## 3 apr 185 1 330 1 failure no
## 4 jun 199 4 –1 0 unknown no
## 5 may 226 1 –1 0 unknown no
## 6 feb 141 2 176 3 failure no
```

**tail**(dat)

```
## age job marital education default balance housing loan contact
## 4516 32 services single secondary no 473 yes no cellular
## 4517 33 services married secondary no –333 yes no cellular
## 4518 57 self-employed married tertiary yes –3313 yes yes unknown
## 4519 57 technician married secondary no 295 no no cellular
## 4520 28 blue-collar married secondary no 1137 no no cellular
## 4521 44 entrepreneur single tertiary no 1136 yes yes cellular
## day month duration campaign pdays previous poutcome y
## 4516 7 jul 624 5 –1 0 unknown no
## 4517 30 jul 329 5 –1 0 unknown no
## 4518 9 may 153 1 –1 0 unknown no
## 4519 19 aug 151 11 –1 0 unknown no
## 4520 6 feb 129 4 211 3 other no
## 4521 3 apr 345 2 249 7 other no
```

We see that the dataframe consists of 4521 rows and 17 columns.

### 9.1.3 write.csv() and write.csv2() Functions

A related function is write.csv() and write.csv2() functions:

write.csv() function is used to create csv files that their values are separated with "," and the decimal character is dot "."

write.csv2() function is used to create csv files that their values are separated with ";" and the decimal character is comma ","

We will use write.csv() function to create another file called bank2.csv and confirm that with list.files(). The function take the first argument the dataframe to create csv file (dat) and the second argument the file name (with its path) as a string ("E:/test/bank2.csv").

**write.csv**(dat, file = "E:/test/bank2.csv")

**list.files**("E:/test")

## [1] "bank.csv"   "bank2.csv"

## [3] "CTG (1).xls"   "CTG.xls"

## [5] "docword.nytimes.txt.gz"   "Online Retail.xlsx"

## [7] "Online.xlsx"   "readme.txt"

## [9] "Residential-Building-Data-Set.xlsx" "Unconfirmed 705073.crdownload"

## [11] "USCensus1990.data.txt"   "vocab.nytimes.txt"

## [13] "vocab.pubmed.txt"

## 9.2 IMPORTING EXCEL FILES (.XLX, .XLSX EXTENSIONS) INTO R

### 9.2.1 read_excel(), read_xls(), and read_xlsx() Functions

package readxl contains several functions for reading Excel files, read_excel(), read_xls(), and read_xlsx() functions read Excel files with the same arguments.

read_xls() and read_xlsx() functions read xls and xlsx files, respectively, while read_excel() calls excel_format() to determine if path is xls or xlsx, based on the file extension and the file itself. Therefore, it is better to use read_xls() or read_xlsx() if you know the file extension.

### 9.2.2 Example

GTC is a data set that consists of measurements of fetal heart rate (FHR) and uterine contraction (UC) features on cardiotocograms classified by expert obstetricians and can be downloaded from:

http://archive.ics.uci.edu/ml/machine-learning-databases/00193/CTG.xls

*Reference:* Ayres et al. (2000).

As we can see, the data has xls extension so it can be read with read_xls() function. Of course, you can open the data in Excel to see how it looks like. The data has 3 sheets, description, data, and raw data. From the arguments of read_xls() ( and also read_xlsx or read_excel), we can control which sheet to read and which columns or rows.

Download the data in your working folder. For example, I put it the computer E compartment and in a folder called test. To confirm that your Excel file is in test folder, run the function, list.files("E:/test"). The functions read_excel(), read_xls(), and read_xlsx() read the first sheet by default and produce a dataframe as an output.

```
library(readxl)

# to locate your file

list.files("E:/test")
## [1] "bank.csv"      "bank2.csv"
## [3] "CTG (1).xls"     "CTG.xls"
## [5] "docword.nytimes.txt.gz"   "Online Retail.xlsx"
## [7] "Online.xlsx"     "readme.txt"
## [9] "Residential-Building-Data-Set.xlsx" "Unconfirmed 705073.
crdownload"
## [11] "USCensus1990.data.txt"   "vocab.nytimes.txt"
## [13] "vocab.pubmed.txt"
# you will see our data, Online Retail.xlsx

# to read your file into R

dat<-read_xls("E:/test/CTG.xls")
## New names:
## * `` -> ...1
## * `` -> ...2
## * `` -> ...4
## * `` -> ...5
## * `` -> ...6
```

```
## * … and 8 more problems
## the first sheet is a data description

class(dat)
## [1] "tbl_df" "tbl" "data.frame"
## third sheet is raw data

dat<-read_xls("E:/test/CTG.xls", sheet = 3)

str(dat)
## Classes 'tbl_df', 'tbl' and 'data.frame': 2130 obs. of 40 variables:
## $ FileName: chr NA "Variab10.txt" "Fmcs_1.txt" "Fmcs_1.txt" ...
## $ Date : POSIXct, format: NA "1996-12-01" ...
## $ SegFile : chr NA "CTG0001.txt" "CTG0002.txt" "CTG0003.txt" ...
## $ b  : num NA 240 5 177 411 533 0 240 62 120 ...
## $ e  : num NA 357 632 779 1192 ...
## $ LBE  : num NA 120 132 133 134 132 134 134 122 122 ...
## $ LB  : num NA 120 132 133 134 132 134 134 122 122 ...
## $ AC  : num NA 0 4 2 2 4 1 1 0 0 ...
## $ FM  : num NA 0 0 0 0 0 0 0 0 0 ...
## $ UC  : num NA 0 4 5 6 5 10 9 0 1 ...
## $ ASTV : num NA 73 17 16 16 16 26 29 83 84 ...
## $ MSTV : num NA 0.5 2.1 2.1 2.4 2.4 5.9 6.3 0.5 0.5 ...
## $ ALTV : num NA 43 0 0 0 0 0 0 6 5 ...
## $ MLTV : num NA 2.4 10.4 13.4 23 19.9 0 0 15.6 13.6 ...
## $ DL  : num NA 0 2 2 2 0 9 6 0 0 ...
## $ DS  : num NA 0 0 0 0 0 0 0 0 0 ...
## $ DP  : num NA 0 0 0 0 0 2 2 0 0 ...
## $ DR  : num NA 0 0 0 0 0 0 0 0 0 ...
## $ Width : num NA 64 130 130 117 117 150 150 68 68 ...
## $ Min  : num NA 62 68 68 53 53 50 50 62 62 ...
```

```
## $ Max  : num NA 126 198 198 170 170 200 200 130 130 ...
## $ Nmax : num NA 2 6 5 11 9 5 6 0 0 ...
## $ Nzeros : num NA 0 1 1 0 0 3 3 0 0 ...
## $ Mode : num NA 120 141 141 137 137 76 71 122 122 ...
## $ Mean : num NA 137 136 135 134 136 107 107 122 122 ...
## $ Median : num NA 121 140 138 137 138 107 106 123 123 ...
## $ Variance: num NA 73 12 13 13 11 170 215 3 3 ...
## $ Tendency: num NA 1 0 0 1 1 0 0 1 1 ...
## $ A  : num NA 0 0 0 0 0 0 0 0 0 ...
## $ B  : num NA 0 0 0 0 1 0 0 0 0 ...
## $ C  : num NA 0 0 0 0 0 0 0 0 0 ...
## $ D  : num NA 0 0 0 0 0 0 0 0 0 ...
## $ E  : num NA 0 0 0 0 0 0 0 0 0 ...
## $ AD  : num NA 0 1 1 1 0 0 0 0 0 ...
## $ DE  : num NA 0 0 0 0 0 0 0 0 0 ...
## $ LD  : num NA 0 0 0 0 0 1 1 0 0 ...
## $ FS  : num NA 1 0 0 0 0 0 0 1 1 ...
## $ SUSP : num NA 0 0 0 0 0 0 0 0 0 ...
## $ CLASS : num NA 9 6 6 6 2 8 8 9 9 ...
## $ NSP  : num NA 2 1 1 1 1 3 3 3 3 ...
head(dat)
## # A tibble: 6 x 40
## FileName Date    SegFile  b  e LBE LB AC FM UC
## <chr> <dttm>    <chr> <dbl> <dbl> <dbl> <dbl> <dbl> <dbl> <dbl>
## 1 <NA>  NA     <NA>  NA NA NA NA NA NA NA
## 2 Variab1… 1996-12-01 00:00:00 CTG000… 240 357 120 120  0  0  0
## 3 Fmcs_1…. 1996-05-03 00:00:00 CTG000…  5 632 132 132  4  0  4
## 4 Fmcs_1…. 1996-05-03 00:00:00 CTG000… 177 779 133 133  2  0  5
## 5 Fmcs_1…. 1996-05-03 00:00:00 CTG000… 411 1192 134 134  2  0  6
## 6 Fmcs_1…. 1996-05-03 00:00:00 CTG000… 533 1147 132 132  4  0  5
## # … with 30 more variables: ASTV <dbl>, MSTV <dbl>, ALTV <dbl>,
```

```
MLTV <dbl>,
## # DL <dbl>, DS <dbl>, DP <dbl>, DR <dbl>, Width <dbl>, Min <dbl>, Max <dbl>,
## # Nmax <dbl>, Nzeros <dbl>, Mode <dbl>, Mean <dbl>, Median <dbl>,
## # Variance <dbl>, Tendency <dbl>, A <dbl>, B <dbl>, C <dbl>, D <dbl>,
## # E <dbl>, AD <dbl>, DE <dbl>, LD <dbl>, FS <dbl>, SUSP <dbl>, CLASS <dbl>,
## # NSP <dbl>
tail(dat)
## # A tibble: 6 x 40
## FileName Date    SegFile  b  e LBE LB AC FM UC
## <chr> <dttm>    <chr> <dbl> <dbl> <dbl> <dbl> <dbl> <dbl> <dbl>
## 1 S800104… 1998-06-06 00:00:00 CTG212… 1576 2596 140 140  1  0 7
## 2 S800104… 1998-06-06 00:00:00 CTG212… 1576 3049 140 140  1  0 9
## 3 S800104… 1998-06-06 00:00:00 CTG212… 2796 3415 142 142  1  1 5
## 4 <NA>  NA    <NA>  NA NA NA NA NA NA NA
## 5 <NA>  NA    <NA>  NA NA NA NA NA NA NA
## 6 <NA>  NA    <NA>  NA NA NA NA NA 564 23
## # … with 30 more variables: ASTV <dbl>, MSTV <dbl>, ALTV <dbl>, MLTV <dbl>,
## # DL <dbl>, DS <dbl>, DP <dbl>, DR <dbl>, Width <dbl>, Min <dbl>, Max <dbl>,
## # Nmax <dbl>, Nzeros <dbl>, Mode <dbl>, Mean <dbl>, Median <dbl>,
## # Variance <dbl>, Tendency <dbl>, A <dbl>, B <dbl>, C <dbl>, D <dbl>,
## # E <dbl>, AD <dbl>, DE <dbl>, LD <dbl>, FS <dbl>, SUSP <dbl>, CLASS <dbl>,
## # NSP <dbl>
##first row is empty and last 3 rows are also empty except for some summary statistics
```

```
# to remove first row

dat2<-dat[–1,]

head(dat2)
## # A tibble: 6 x 40
## FileName Date    SegFile  b  e LBE LB AC FM UC
## <chr> <dttm>    <chr> <dbl> <dbl> <dbl> <dbl> <dbl> <dbl> <dbl>
## 1 Variab1… 1996-12-01 00:00:00 CTG000… 240 357 120 120  0  0  0
## 2 Fmcs_1…. 1996-05-03 00:00:00 CTG000…  5 632 132 132  4  0  4
## 3 Fmcs_1…. 1996-05-03 00:00:00 CTG000… 177 779 133 133  2  0  5
## 4 Fmcs_1…. 1996-05-03 00:00:00 CTG000… 411 1192 134 134  2  0  6
## 5 Fmcs_1…. 1996-05-03 00:00:00 CTG000… 533 1147 132 132  4  0  5
## 6 Fmcs_2…. 1996-05-03 00:00:00 CTG000…  0 953 134 134  1  0 10
## # … with 30 more variables: ASTV <dbl>, MSTV <dbl>, ALTV <dbl>,
MLTV <dbl>,
## # DL <dbl>, DS <dbl>, DP <dbl>, DR <dbl>, Width <dbl>, Min <dbl>,
Max <dbl>,
## # Nmax <dbl>, Nzeros <dbl>, Mode <dbl>, Mean <dbl>, Median <dbl>,
## # Variance <dbl>, Tendency <dbl>, A <dbl>, B <dbl>, C <dbl>, D <dbl>,
## # E <dbl>, AD <dbl>, DE <dbl>, LD <dbl>, FS <dbl>, SUSP <dbl>,
CLASS <dbl>,
## # NSP <dbl>
tail(dat2)
## # A tibble: 6 x 40
## FileName Date    SegFile  b  e LBE LB AC FM UC
## <chr> <dttm>    <chr> <dbl> <dbl> <dbl> <dbl> <dbl> <dbl> <dbl>
## 1 S800104… 1998-06-06 00:00:00 CTG212… 1576 2596 140 140  1  0
7
## 2 S800104… 1998-06-06 00:00:00 CTG212… 1576 3049 140 140  1  0
9
```

```
## 3 S800104… 1998-06-06 00:00:00 CTG212… 2796 3415 142 142 1 1 5
## 4 <NA> NA <NA> NA NA NA NA NA NA NA
## 5 <NA> NA <NA> NA NA NA NA NA NA NA
## 6 <NA> NA <NA> NA NA NA NA NA 564 23
## # … with 30 more variables: ASTV <dbl>, MSTV <dbl>, ALTV <dbl>, MLTV <dbl>,
## # DL <dbl>, DS <dbl>, DP <dbl>, DR <dbl>, Width <dbl>, Min <dbl>, Max <dbl>,
## # Nmax <dbl>, Nzeros <dbl>, Mode <dbl>, Mean <dbl>, Median <dbl>,
## # Variance <dbl>, Tendency <dbl>, A <dbl>, B <dbl>, C <dbl>, D <dbl>,
## # E <dbl>, AD <dbl>, DE <dbl>, LD <dbl>, FS <dbl>, SUSP <dbl>, CLASS <dbl>,
## # NSP <dbl>
str(dat2)
## Classes ‘tbl_df’, ‘tbl’ and ‘data.frame’: 2129 obs. of 40 variables:
## $ FileName: chr “Variab10.txt” “Fmcs_1.txt” “Fmcs_1.txt” “Fmcs_1.txt” ...
## $ Date : POSIXct, format: “1996-12-01” “1996-05-03” ...
## $ SegFile : chr “CTG0001.txt” “CTG0002.txt” “CTG0003.txt” “CTG0004.txt” ...
## $ b : num 240 5 177 411 533 0 240 62 120 181 ...
## $ e : num 357 632 779 1192 1147 ...
## $ LBE : num 120 132 133 134 132 134 134 122 122 122 ...
## $ LB : num 120 132 133 134 132 134 134 122 122 122 ...
## $ AC : num 0 4 2 2 4 1 1 0 0 0 ...
## $ FM : num 0 0 0 0 0 0 0 0 0 0 ...
## $ UC : num 0 4 5 6 5 10 9 0 1 3 ...
## $ ASTV : num 73 17 16 16 16 26 29 83 84 86 ...
## $ MSTV : num 0.5 2.1 2.1 2.4 2.4 5.9 6.3 0.5 0.5 0.3 ...
## $ ALTV : num 43 0 0 0 0 0 0 6 5 6 ...
## $ MLTV : num 2.4 10.4 13.4 23 19.9 0 0 15.6 13.6 10.6 ...
```

```
## $ DL  : num 0 2 2 2 0 9 6 0 0 0 ...
## $ DS  : num 0 0 0 0 0 0 0 0 0 0 ...
## $ DP  : num 0 0 0 0 0 2 2 0 0 0 ...
## $ DR  : num 0 0 0 0 0 0 0 0 0 0 ...
## $ Width : num 64 130 130 117 117 150 150 68 68 68 ...
## $ Min  : num 62 68 68 53 53 50 50 62 62 62 ...
## $ Max  : num 126 198 198 170 170 200 200 130 130 130 ...
## $ Nmax : num 2 6 5 11 9 5 6 0 0 1 ...
## $ Nzeros : num 0 1 1 0 0 3 3 0 0 0 ...
## $ Mode : num 120 141 141 137 137 76 71 122 122 122 ...
## $ Mean : num 137 136 135 134 136 107 107 122 122 122 ...
## $ Median : num 121 140 138 137 138 107 106 123 123 123 ...
## $ Variance: num 73 12 13 13 11 170 215 3 3 1 ...
## $ Tendency: num 1 0 0 1 1 0 0 1 1 1 ...
## $ A  : num 0 0 0 0 0 0 0 0 0 0 ...
## $ B  : num 0 0 0 0 1 0 0 0 0 0 ...
## $ C  : num 0 0 0 0 0 0 0 0 0 0 ...
## $ D  : num 0 0 0 0 0 0 0 0 0 0 ...
## $ E  : num 0 0 0 0 0 0 0 0 0 0 ...
## $ AD  : num 0 1 1 1 0 0 0 0 0 0 ...
## $ DE  : num 0 0 0 0 0 0 0 0 0 0 ...
## $ LD  : num 0 0 0 0 0 1 1 0 0 0 ...
## $ FS  : num 1 0 0 0 0 0 0 1 1 1 ...
## $ SUSP : num 0 0 0 0 0 0 0 0 0 0 ...
## $ CLASS : num 9 6 6 6 2 8 8 9 9 9 ...
## $ NSP  : num 2 1 1 1 1 3 3 3 3 3 ...
# to remove last three rows

dat3<-dat2[1:2126,]

head(dat3)
```

```
## # A tibble: 6 x 40
## FileName Date    SegFile  b  e LBE LB AC FM UC
## <chr> <dttm>    <chr> <dbl> <dbl> <dbl> <dbl> <dbl> <dbl> <dbl>
## 1 Variabl… 1996-12-01 00:00:00 CTG000… 240 357 120 120  0  0  0
## 2 Fmcs_1…. 1996-05-03 00:00:00 CTG000…  5 632 132 132  4  0  4
## 3 Fmcs_1…. 1996-05-03 00:00:00 CTG000… 177 779 133 133  2  0  5
## 4 Fmcs_1…. 1996-05-03 00:00:00 CTG000… 411 1192 134 134  2  0  6
## 5 Fmcs_1…. 1996-05-03 00:00:00 CTG000… 533 1147 132 132  4  0  5
## 6 Fmcs_2…. 1996-05-03 00:00:00 CTG000…  0 953 134 134  1  0 10
## # … with 30 more variables: ASTV <dbl>, MSTV <dbl>, ALTV <dbl>,
MLTV <dbl>,
## # DL <dbl>, DS <dbl>, DP <dbl>, DR <dbl>, Width <dbl>, Min <dbl>,
Max <dbl>,
## # Nmax <dbl>, Nzeros <dbl>, Mode <dbl>, Mean <dbl>, Median <dbl>,
## # Variance <dbl>, Tendency <dbl>, A <dbl>, B <dbl>, C <dbl>, D <dbl>,
## # E <dbl>, AD <dbl>, DE <dbl>, LD <dbl>, FS <dbl>, SUSP <dbl>,
CLASS <dbl>,
## # NSP <dbl>
tail(dat3)
## # A tibble: 6 x 40
## FileName Date    SegFile  b  e LBE LB AC FM UC
## <chr> <dttm>    <chr> <dbl> <dbl> <dbl> <dbl> <dbl> <dbl> <dbl>
## 1 S800104… 1998-06-06 00:00:00 CTG212… 1143 1947 140 140  0  0
4
## 2 S800104… 1998-06-06 00:00:00 CTG212… 2059 2867 140 140  0  0
6
## 3 S800104… 1998-06-06 00:00:00 CTG212… 1576 2867 140 140  1  0
9
## 4 S800104… 1998-06-06 00:00:00 CTG212… 1576 2596 140 140  1  0
7
## 5 S800104… 1998-06-06 00:00:00 CTG212… 1576 3049 140 140  1  0
9
## 6 S800104… 1998-06-06 00:00:00 CTG212… 2796 3415 142 142  1  1
```

```
5
## # … with 30 more variables: ASTV <dbl>, MSTV <dbl>, ALTV <dbl>, MLTV <dbl>,
## # DL <dbl>, DS <dbl>, DP <dbl>, DR <dbl>, Width <dbl>, Min <dbl>, Max <dbl>,
## # Nmax <dbl>, Nzeros <dbl>, Mode <dbl>, Mean <dbl>, Median <dbl>,
## # Variance <dbl>, Tendency <dbl>, A <dbl>, B <dbl>, C <dbl>, D <dbl>,
## # E <dbl>, AD <dbl>, DE <dbl>, LD <dbl>, FS <dbl>, SUSP <dbl>, CLASS <dbl>,
## # NSP <dbl>
str(dat3)
## Classes 'tbl_df', 'tbl' and 'data.frame': 2126 obs. of 40 variables:
## $ FileName: chr "Variab10.txt" "Fmcs_1.txt" "Fmcs_1.txt" "Fmcs_1.txt" ...
## $ Date : POSIXct, format: "1996-12-01" "1996-05-03" ...
## $ SegFile : chr "CTG0001.txt" "CTG0002.txt" "CTG0003.txt" "CTG0004.txt" ...
## $ b : num 240 5 177 411 533 0 240 62 120 181 ...
## $ e : num 357 632 779 1192 1147 ...
## $ LBE : num 120 132 133 134 132 134 134 122 122 122 ...
## $ LB : num 120 132 133 134 132 134 134 122 122 122 ...
## $ AC : num 0 4 2 2 4 1 1 0 0 0 ...
## $ FM : num 0 0 0 0 0 0 0 0 0 0 ...
## $ UC : num 0 4 5 6 5 10 9 0 1 3 ...
## $ ASTV : num 73 17 16 16 16 26 29 83 84 86 ...
## $ MSTV : num 0.5 2.1 2.1 2.4 2.4 5.9 6.3 0.5 0.5 0.3 ...
## $ ALTV : num 43 0 0 0 0 0 0 6 5 6 ...
## $ MLTV : num 2.4 10.4 13.4 23 19.9 0 0 15.6 13.6 10.6 ...
## $ DL : num 0 2 2 2 0 9 6 0 0 0 ...
## $ DS : num 0 0 0 0 0 0 0 0 0 0 ...
## $ DP : num 0 0 0 0 0 2 2 0 0 0 ...
## $ DR : num 0 0 0 0 0 0 0 0 0 0 ...
```

```
## $ Width : num 64 130 130 117 117 150 150 68 68 68 ...
## $ Min  : num 62 68 68 53 53 50 50 62 62 62 ...
## $ Max  : num 126 198 198 170 170 200 200 130 130 130 ...
## $ Nmax : num 2 6 5 11 9 5 6 0 0 1 ...
## $ Nzeros : num 0 1 1 0 0 3 3 0 0 0 ...
## $ Mode : num 120 141 141 137 137 76 71 122 122 122 ...
## $ Mean : num 137 136 135 134 136 107 107 122 122 122 ...
## $ Median : num 121 140 138 137 138 107 106 123 123 123 ...
## $ Variance: num 73 12 13 13 11 170 215 3 3 1 ...
## $ Tendency: num 1 0 0 1 1 0 0 1 1 1 ...
## $ A  : num 0 0 0 0 0 0 0 0 0 0 ...
## $ B  : num 0 0 0 0 1 0 0 0 0 0 ...
## $ C  : num 0 0 0 0 0 0 0 0 0 0 ...
## $ D  : num 0 0 0 0 0 0 0 0 0 0 ...
## $ E  : num 0 0 0 0 0 0 0 0 0 0 ...
## $ AD  : num 0 1 1 1 0 0 0 0 0 0 ...
## $ DE  : num 0 0 0 0 0 0 0 0 0 0 ...
## $ LD  : num 0 0 0 0 0 1 1 0 0 0 ...
## $ FS  : num 1 0 0 0 0 0 0 1 1 1 ...
## $ SUSP : num 0 0 0 0 0 0 0 0 0 0 ...
## $ CLASS : num 9 6 6 6 2 8 8 9 9 9 ...
## $ NSP  : num 2 1 1 1 1 3 3 3 3 3 ...
```

So the final dataframe consists of 2126 rows and 40 columns. According to data web site, the important 23 columns are LB, AC, FM, UC, DL, DS, DP, ASTV, MSTV, ALTV, MLTV, Width, Min, Max, Nmax, Nzeros, Mode, Mean, Median, Variance, Tendency, CLASS, NSP which have column numbers 7:17,19:28,39,40.

```
# to extract important columns

dat4<-dat3[,c(7:17,19:28,39,40)]
```

```
names(dat4)
## [1] "LB" "AC" "FM" "UC" "ASTV" "MSTV"
## [7] "ALTV" "MLTV" "DL" "DS" "DP" "Width"
## [13] "Min" "Max" "Nmax" "Nzeros" "Mode" "Mean"
## [19] "Median" "Variance" "Tendency" "CLASS" "NSP"
head(dat4)
## # A tibble: 6 x 23
## LB AC FM UC ASTV MSTV ALTV MLTV DL DS DP Width Min
## <dbl> <dbl> <dbl> <dbl> <dbl> <dbl> <dbl> <dbl> <dbl> <dbl> <dbl>
<dbl> <dbl>
## 1 120 0 0 0 73 0.5 43 2.4 0 0 0 64 62
## 2 132 4 0 4 17 2.1 0 10.4 2 0 0 130 68
## 3 133 2 0 5 16 2.1 0 13.4 2 0 0 130 68
## 4 134 2 0 6 16 2.4 0 23 2 0 0 117 53
## 5 132 4 0 5 16 2.4 0 19.9 0 0 0 117 53
## 6 134 1 0 10 26 5.9 0 0 9 0 2 150 50
## # … with 10 more variables: Max <dbl>, Nmax <dbl>, Nzeros <dbl>,
Mode <dbl>,
## # Mean <dbl>, Median <dbl>, Variance <dbl>, Tendency <dbl>, CLASS
<dbl>,
## # NSP <dbl>
tail(dat4)
## # A tibble: 6 x 23
## LB AC FM UC ASTV MSTV ALTV MLTV DL DS DP Width Min
## <dbl> <dbl> <dbl> <dbl> <dbl> <dbl> <dbl> <dbl> <dbl> <dbl> <dbl>
<dbl> <dbl>
## 1 140 0 0 4 77 0.7 17 6 1 0 0 31 124
## 2 140 0 0 6 79 0.2 25 7.2 0 0 0 40 137
## 3 140 1 0 9 78 0.4 22 7.1 0 0 0 66 103
## 4 140 1 0 7 79 0.4 20 6.1 0 0 0 67 103
## 5 140 1 0 9 78 0.4 27 7 0 0 0 66 103
```

```
## 6 142  1  1  5 74 0.4 36 5  0  0  0 42 117
## # … with 10 more variables: Max <dbl>, Nmax <dbl>, Nzeros <dbl>, Mode <dbl>,
## # Mean <dbl>, Median <dbl>, Variance <dbl>, Tendency <dbl>, CLASS <dbl>,
## # NSP <dbl>
```

**str**(dat4)

```
## Classes 'tbl_df', 'tbl' and 'data.frame': 2126 obs. of 23 variables:
## $ LB  : num 120 132 133 134 132 134 134 122 122 122 ...
## $ AC  : num 0 4 2 2 4 1 1 0 0 0 ...
## $ FM  : num 0 0 0 0 0 0 0 0 0 0 ...
## $ UC  : num 0 4 5 6 5 10 9 0 1 3 ...
## $ ASTV : num 73 17 16 16 16 26 29 83 84 86 ...
## $ MSTV : num 0.5 2.1 2.1 2.4 2.4 5.9 6.3 0.5 0.5 0.3 ...
## $ ALTV : num 43 0 0 0 0 0 0 6 5 6 ...
## $ MLTV : num 2.4 10.4 13.4 23 19.9 0 0 15.6 13.6 10.6 ...
## $ DL  : num 0 2 2 2 0 9 6 0 0 0 ...
## $ DS  : num 0 0 0 0 0 0 0 0 0 0 ...
## $ DP  : num 0 0 0 0 0 2 2 0 0 0 ...
## $ Width : num 64 130 130 117 117 150 150 68 68 68 ...
## $ Min  : num 62 68 68 53 53 50 50 62 62 62 ...
## $ Max  : num 126 198 198 170 170 200 200 130 130 130 ...
## $ Nmax : num 2 6 5 11 9 5 6 0 0 1 ...
## $ Nzeros : num 0 1 1 0 0 3 3 0 0 0 ...
## $ Mode : num 120 141 141 137 137 76 71 122 122 122 ...
## $ Mean : num 137 136 135 134 136 107 107 122 122 122 ...
## $ Median : num 121 140 138 137 138 107 106 123 123 123 ...
## $ Variance: num 73 12 13 13 11 170 215 3 3 1 ...
## $ Tendency: num 1 0 0 1 1 0 0 1 1 1 ...
## $ CLASS : num 9 6 6 6 2 8 8 9 9 9 ...
## $ NSP  : num 2 1 1 1 1 3 3 3 3 3 ...
```

So final dataframe consists of 2126 rows and 23 columns.

## 9.3 IMPORTING TAB SEPARATED FILES (.TXT EXTENSION) INTO R

### 9.3.1 read.table(), read.delim(), or read.delim2() Functions

If your files are tab delimited files or with .txt extension, these files can be read with read.table(), read.delim(), or read.delim2() functions.

The default behavior of read.table() function is for files that their values are separated with "", the decimal character is dot ".", and there is no column names row as first row (header = FALSE).

The default behavior of read.delim() function is for files that their values are separated with "\t", the decimal character is dot ".", and there are column names row as first row (header = TRUE).

The default behavior of read.delim2() function is for files that their values are separated with "\t", the decimal character is comma ",", and there are column names row as first row (header = TRUE).

### 9.3.2 Example: Bag of Words for NYTimes News Articles

Go to http://archive.ics.uci.edu/ml/machine-learning-databases/bag-of-words/ and download vocab.nytimes.txt file in your working directory. This is the vocabulary file.

*Sources:* ldc.upenn.edu

For example, I put this data in the computer E compartment and in a folder called test. To confirm that your vocab.nytimes.txt file is in test folder, run the function, list.files("E:/test").

Open the file using WordPad or Notepad to see how the values are separated. You will see that the values are not separated (sep="") and there is no header or column names. However, this file can be read with read.table(), read.delim(), or read.delim2() functions by manipulating its arguments

```
# to locate your file

list.files("E:/test")
## [1] "bank.csv"     "bank2.csv"
```

```
## [3] "CTG (1).xls"     "CTG.xls"
## [5] "docword.nytimes.txt.gz"   "Online Retail.xlsx"
## [7] "Online.xlsx"     "readme.txt"
## [9] "Residential-Building-Data-Set.xlsx" "Unconfirmed 705073.crdownload"
## [11] "USCensus1990.data.txt"   "vocab.nytimes.txt"
## [13] "vocab.pubmed.txt"
# to read your file into R

dat<-read.table("E:/test/vocab.nytimes.txt")

dat2<-read.delim("E:/test/vocab.nytimes.txt", sep = """, dec = ".", header = FALSE)

dat3<-read.delim2("E:/test/vocab.nytimes.txt", sep = """, dec = ".", header = FALSE)

identical(dat, dat2)
## [1] TRUE
identical(dat,dat3)
## [1] TRUE
identical(dat2,dat3)
## [1] TRUE
```

The three functions have read the same file and produced the same dataframe as evident from the identical function.

The produced dataset is a dataframe and so can be explored by different functions from the previous lessons.

```
class(dat)
## [1] "data.frame"
str(dat)
```

```
## 'data.frame': 102660 obs. of 1 variable:
## $ V1: Factor w/ 102660 levels "aah","aahed",..: 1 2 3 4 5 6 7 8 9 10 ...
head(dat)
##  V1
## 1 aah
## 2 aahed
## 3 aaron
## 4 aback
## 5 abacus
## 6 abajo
# to see more vocabulary values, set a larger value to head() function

# to see first 30 rows

head(dat,30)
##    V1
## 1   aah
## 2   aahed
## 3   aaron
## 4   aback
## 5  abacus
## 6   abajo
## 7  abalone
## 8  abandon
## 9  abandoned
## 10 abandoning
## 11 abandonment
## 12  abandono
## 13  abarnard
## 14  abashed
## 15  abate
```

```
## 16  abated
## 17 abatement
## 18  abating
## 19  abbey
## 20  abbot
## 21 abbreviated
## 22 abbreviation
## 23   abc
## 24  abcnew
## 25  abdicate
## 26 abdicated
## 27 abdicating
## 28 abdication
## 29  abdomen
## 30 abdominal
```

```
tail(dat)
##       V1
## 102655    zzz_zydeco
## 102656 zzz_zydrunas_ilgauskas
## 102657   zzz_zyprexa
## 102658    zzz_zyrtec
## 102659    zzz_zz_top
## 102660    zzz_zzzt
```

We see that the dataframe consists of 102,660 rows and 1 column.

### 9.3.3 Example 2: US Census Data (1990) Data Set

Go to http://archive.ics.uci.edu/ml/machine-learning-databases/census1990-mld/ and download USCensus1990.data.txt file in your working directory. This data is a random sample of the USCensus1990raw data set that contains a one percent sample of the public use microdata samples (PUMS) person records drawn from the full 1990 census sample.

[*Source:* The USCensus1990raw data set was obtained from the (U.S. Department of Commerce) Census Bureau website using the Data Extraction

System. This system can be found at: http://dataferrett.census.gov/.]

For example, I put this data in the computer E compartment and in a folder called test. To confirm that your USCensus1990.data.txt file is in test folder, run the function, list.files("E:/test").

Open the file using WordPad or Notepad to see how the values are separated. You will see that the values are separated with a comma (sep=",") and the first row is a column names. However, this file can be read with read.table(), read.delim(), read.delim2(), or read.csv() functions by manipulating its arguments

Because the dataset has many rows, by specifying the argument, nrows = 10000, we will read the first 10000 rows of this data into the resulting dataframe.

```
# to locate your file

list.files("E:/test")
## [1] "bank.csv"     "bank2.csv"
## [3] "CTG (1).xls"     "CTG.xls"
## [5] "docword.nytimes.txt.gz"   "Online Retail.xlsx"
## [7] "Online.xlsx"     "readme.txt"
## [9] "Residential-Building-Data-Set.xlsx" "Unconfirmed 705073.crdownload"
## [11] "USCensus1990.data.txt"   "vocab.nytimes.txt"
## [13] "vocab.pubmed.txt"
# to read your file into R

dat<-read.table("E:/test/USCensus1990.data.txt", header = TRUE, sep = ",", dec = ".", nrows = 10000)

dat2<-read.delim("E:/test/USCensus1990.data.txt", sep = ",", dec = ".", header = TRUE, nrows = 10000)

dat3<-read.delim2("E:/test/USCensus1990.data.txt", sep = ",", dec = ".", header = TRUE, nrows = 10000)
```

dat4<-**read.csv**("E:/test/USCensus1990.data.txt", sep = ",", dec = ".", header = TRUE, nrows = 10000)

**identical**(dat, dat2)

## [1] TRUE

**identical**(dat,dat3)

## [1] TRUE

**identical**(dat,dat4)

## [1] TRUE

**identical**(dat2,dat3)

## [1] TRUE

**identical**(dat2,dat4)

## [1] TRUE

**identical**(dat3,dat4)

## [1] TRUE

The four functions have read the same file and produced the same dataframe as evident from the identical function.

The produced dataset is a dataframe and so can be explored by different functions from the previous lessons

**class**(dat)

## [1] "data.frame"

**str**(dat)

## 'data.frame': 10000 obs. of 69 variables:

## $ caseid : int 10000 10001 10002 10003 10004 10005 10006 10007 10008 10009 ...

## $ dAge : int 5 6 3 4 7 1 1 4 6 3 ...

## $ dAncstry1: int 0 1 1 1 1 1 1 1 1 1 ...

## $ dAncstry2: int 1 1 2 2 1 2 1 2 1 12 ...

```
## $ iAvail : int 0 0 0 0 0 0 0 0 0 0 ...
## $ iCitizen : int 0 0 0 0 0 0 0 0 0 0 ...
## $ iClass : int 5 7 7 1 0 0 0 6 1 1 ...
## $ dDepart : int 3 5 4 3 0 0 0 0 0 0 ...
## $ iDisabl1 : int 2 2 2 2 2 0 0 2 2 2 ...
## $ iDisabl2 : int 2 2 2 2 2 0 0 2 2 2 ...
## $ iEnglish : int 1 0 0 0 0 0 0 0 0 0 ...
## $ iFeb55 : int 0 0 0 0 0 0 0 0 0 0 ...
## $ iFertil : int 1 3 1 3 3 0 0 4 7 0 ...
## $ dHispanic: int 0 0 0 0 0 0 0 0 0 0 ...
## $ dHour89 : int 4 1 4 3 0 0 0 5 1 3 ...
## $ dHours : int 3 1 4 3 0 0 0 5 0 0 ...
## $ iImmigr : int 0 0 0 0 0 0 0 0 0 0 ...
## $ dIncome1 : int 2 1 1 1 0 0 0 2 1 1 ...
## $ dIncome2 : int 0 0 0 0 0 0 0 1 0 0 ...
## $ dIncome3 : int 0 0 1 0 0 0 0 0 0 0 ...
## $ dIncome4 : int 1 0 0 0 0 0 0 0 0 0 ...
## $ dIncome5 : int 0 0 0 0 1 0 0 0 0 0 ...
## $ dIncome6 : int 0 1 0 0 0 0 0 0 0 0 ...
## $ dIncome7 : int 0 0 0 0 0 0 0 0 0 0 ...
## $ dIncome8 : int 0 0 0 1 0 0 0 0 0 0 ...
## $ dIndustry: int 10 4 1 4 0 0 0 9 7 4 ...
## $ iKorean : int 0 0 0 0 0 0 0 0 0 0 ...
## $ iLang1 : int 1 2 2 2 2 2 2 2 2 2 ...
## $ iLooking : int 0 0 0 0 2 0 0 0 2 2 ...
## $ iMarital : int 1 0 4 2 0 4 4 0 0 4 ...
## $ iMay75880: int 0 0 0 0 0 0 0 0 0 0 ...
## $ iMeans : int 1 1 10 1 0 0 0 11 0 0 ...
## $ iMilitary: int 4 4 4 4 4 0 0 4 4 2 ...
## $ iMobility: int 2 1 1 1 1 1 1 1 1 1 ...
## $ iMobillim: int 2 2 2 2 2 0 0 2 2 2 ...
```

```
## $ dOccup : int 3 2 4 2 0 0 0 3 6 6 ...
## $ iOthrserv: int 0 0 0 0 0 0 0 0 0 0 ...
## $ iPerscare: int 2 2 2 2 2 0 0 2 2 2 ...
## $ dPOB  : int 0 0 0 0 0 0 0 0 0 0 ...
## $ dPoverty : int 2 2 2 2 2 2 2 2 2 2 ...
## $ dPwgt1 : int 1 2 1 1 1 0 1 1 1 1 ...
## $ iRagechld: int 4 4 4 2 4 4 0 2 2 0 ...
## $ dRearning: int 3 2 2 2 0 0 0 3 2 2 ...
## $ iRelat1 : int 0 1 2 0 1 2 2 1 1 2 ...
## $ iRelat2 : int 0 0 0 0 0 0 0 0 0 0 ...
## $ iRemplpar: int 0 0 0 0 0 121 121 0 0 0 ...
## $ iRiders : int 3 1 0 1 0 0 0 0 0 0 ...
## $ iRlabor : int 1 1 1 1 6 0 0 1 6 3 ...
## $ iRownchld: int 0 0 0 0 0 1 1 0 0 0 ...
## $ dRpincome: int 3 2 2 2 2 0 0 3 2 2 ...
## $ iRPOB : int 22 10 10 10 22 10 10 10 10 10 ...
## $ iRrelchld: int 0 0 0 0 0 1 1 0 0 0 ...
## $ iRspouse : int 3 1 6 4 1 0 0 1 1 6 ...
## $ iRvetserv: int 0 0 0 0 0 0 0 0 0 1 ...
## $ iSchool : int 1 1 1 1 1 2 2 1 1 1 ...
## $ iSept80 : int 0 0 0 0 0 0 0 0 0 1 ...
## $ iSex  : int 1 1 1 1 1 1 0 1 1 0 ...
## $ iSubfam1 : int 0 0 0 0 0 0 0 0 0 0 ...
## $ iSubfam2 : int 0 0 0 0 0 0 0 0 0 0 ...
## $ iTmpabsnt: int 0 0 0 0 3 0 0 0 3 1 ...
## $ dTravtime: int 5 1 2 1 0 0 0 0 0 0 ...
## $ iVietnam : int 0 0 0 0 0 0 0 0 0 0 ...
## $ dWeek89 : int 2 2 2 1 0 0 0 2 1 1 ...
## $ iWork89 : int 1 1 1 1 2 0 0 1 1 1 ...
## $ iWorklwk : int 1 1 1 1 2 0 0 1 2 2 ...
## $ iWWII : int 0 0 0 0 0 0 0 0 0 0 ...
```

```
## $ iYearsch : int 11 5 10 10 5 4 4 11 10 8 ...
## $ iYearwrk : int 1 1 1 1 6 0 0 1 1 1 ...
## $ dYrsserv : int 0 0 0 0 0 0 0 0 0 1 ...
```

**head**(dat)

```
## caseid dAge dAncstry1 dAncstry2 iAvail iCitizen iClass dDepart iDisabl1
## 1 10000 5 0 1 0 0 5 3 2
## 2 10001 6 1 1 0 0 7 5 2
## 3 10002 3 1 2 0 0 7 4 2
## 4 10003 4 1 2 0 0 1 3 2
## 5 10004 7 1 1 0 0 0 0 2
## 6 10005 1 1 2 0 0 0 0 0
## iDisabl2 iEnglish iFeb55 iFertil dHispanic dHour89 dHours iImmigr dIncome1
## 1 2 1 0 1 0 4 3 0 2
## 2 2 0 0 3 0 1 1 0 1
## 3 2 0 0 1 0 4 4 0 1
## 4 2 0 0 3 0 3 3 0 1
## 5 2 0 0 3 0 0 0 0 0
## 6 0 0 0 0 0 0 0 0 0
## dIncome2 dIncome3 dIncome4 dIncome5 dIncome6 dIncome7 dIncome8 dIndustry
## 1 0 0 1 0 0 0 0 10
## 2 0 0 0 0 1 0 0 4
## 3 0 1 0 0 0 0 0 1
## 4 0 0 0 0 0 0 1 4
## 5 0 0 0 1 0 0 0 0
## 6 0 0 0 0 0 0 0 0
## iKorean iLang1 iLooking iMarital iMay75880 iMeans iMilitary iMobility
## 1 0 1 0 1 0 1 4 2
## 2 0 2 0 0 0 1 4 1
## 3 0 2 0 4 0 10 4 1
## 4 0 2 0 2 0 1 4 1
```

```
## 5 0 2 2 0 0 0 4 1
## 6 0 2 0 4 0 0 0 1
## iMobillim dOccup iOthrserv iPerscare dPOB dPoverty dPwgt1 iRagechld dRearning
## 1 2 3 0 2 0 2 1 4 3
## 2 2 2 0 2 0 2 2 4 2
## 3 2 4 0 2 0 2 1 4 2
## 4 2 2 0 2 0 2 1 2 2
## 5 2 0 0 2 0 2 1 4 0
## 6 0 0 0 0 0 2 0 4 0
## iRelat1 iRelat2 iRemplpar iRiders iRlabor iRownchld dRpincome iRPOB iRrelchld
## 1 0 0 0 3 1 0 3 22 0
## 2 1 0 0 1 1 0 2 10 0
## 3 2 0 0 0 1 0 2 10 0
## 4 0 0 0 1 1 0 2 10 0
## 5 1 0 0 0 6 0 2 22 0
## 6 2 0 121 0 0 1 0 10 1
## iRspouse iRvetserv iSchool iSept80 iSex iSubfam1 iSubfam2 iTmpabsnt dTravtime
## 1 3 0 1 0 1 0 0 0 5
## 2 1 0 1 0 1 0 0 0 1
## 3 6 0 1 0 1 0 0 0 2
## 4 4 0 1 0 1 0 0 0 1
## 5 1 0 1 0 1 0 0 3 0
## 6 0 0 2 0 1 0 0 0 0
## iVietnam dWeek89 iWork89 iWorklwk iWWII iYearsch iYearwrk dYrsserv
## 1 0 2 1 1 0 11 1 0
## 2 0 2 1 1 0 5 1 0
```

```
## 3 0 2 1 1 0 10 1 0
## 4 0 1 1 1 0 10 1 0
## 5 0 0 2 2 0 5 6 0
## 6 0 0 0 0 0 4 0 0
```

```
tail(dat)
## caseid dAge dAncstry1 dAncstry2 iAvail iCitizen iClass dDepart iDisabl1
## 9995 19994 3 0 1 0 4 1 3 1
## 9996 19995 1 0 1 0 0 0 0 0
## 9997 19996 5 11 1 0 0 1 4 2
## 9998 19997 3 1 2 0 0 1 3 2
## 9999 19998 3 1 1 0 0 0 0 2
## 10000 19999 1 1 3 0 0 0 0 0
## iDisabl2 iEnglish iFeb55 iFertil dHispanic dHour89 dHours iImmigr
## 9995 2 0 0 0 0 3 3 5
## 9996 0 0 0 0 0 0 0 0
## 9997 2 0 0 2 0 3 5 0
## 9998 2 0 0 0 0 3 3 0
## 9999 2 0 0 5 0 0 0 0
## 10000 0 0 0 0 0 0 0 0
## dIncome1 dIncome2 dIncome3 dIncome4 dIncome5 dIncome6 dIncome7 dIncome8
## 9995 1 0 0 0 0 0 0 0
## 9996 0 0 0 0 0 0 0 0
## 9997 1 0 0 0 0 0 0 0
## 9998 3 0 0 1 0 0 0 0
## 9999 0 0 0 0 0 1 0 0
## 10000 0 0 0 0 0 0 0 0
## dIndustry iKorean iLang1 iLooking iMarital iMay75880 iMeans iMilitary
## 9995 3 0 2 0 4 0 1 4
```

```
## 9996  0 0 2 0 4  0 0  0
## 9997  9 0 2 0 2  0 1  4
## 9998  3 0 2 0 0  0 1  4
## 9999  0 0 2 2 0  0 0  4
## 10000  0 0 0 0 4  0 0  0
##  iMobility iMobillim dOccup iOthrserv iPerscare dPOB dPoverty dPwgt1
## 9995  1  2 5  0  2 4  2  1
## 9996  1  0 0  0  0 0  2  1
## 9997  1  2 1  0  2 0  2  1
## 9998  2  2 1  0  2 0  2  2
## 9999  2  2 0  0  2 0  1  1
## 10000  0  0 0  0  0 0  2  1
##    iRagechld dRearning iRelat1 iRelat2 iRemplpar iRiders iRlabor
iRownchld
## 9995  0  2 2 0  0 1  1  0
## 9996  4  0 2 0 111  0 0  1
## 9997  4  2 0 0  0 1  1  0
## 9998  0  4 0 0  0 1  1  0
## 9999  1  0 1 0  0 0  6  0
## 10000  0  0 2 0 121  0 0  1
##  dRpincome iRPOB iRrelchld iRspouse iRvetserv iSchool iSept80 iSex
## 9995  2 52  0 6  0 1  0 0
## 9996  0 10  1 0  0 2  0 1
## 9997  2 10  0 4  0 1  0 1
## 9998  4 10  0 1  0 1  0 0
## 9999  2 10  0 1  0 1  0 1
## 10000  0 10  1 0  0 0  0 0
##  iSubfam1 iSubfam2 iTmpabsnt dTravtime iVietnam dWeek89 iWork89
iWorklwk
## 9995  0 0  0  4 0 1  1  2
## 9996  0 0  0  0 0 0 0  0
## 9997  0 0  0  2 0 2  1  1
```

```
## 9998  0 0  0  3 0 2 1 1
## 9999  0 0  3  0 0 0 2 2
## 10000 0 0  0  0 0 0 0 0
##  iWWII iYearsch iYearwrk dYrsserv
## 9995  0  7  1  0
## 9996  0  5  0  0
## 9997  0  11  1  0
## 9998  0  14  1  0
## 9999  0  7  7  0
## 10000  0  0  0  0
```

We see that the produced dataframe consists of 10,000 rows and 69 columns.

CHAPTER 10

# BASIC DATA WRANGLING WITH TIDYVERSE

## CONTENTS

## 10.1 TIDY DATASETS

A tidy dataset has the following properties:

- Each variable (feature, characteristic) forms a column;
- Each observation (experimental unit) forms a row.

Most datasets do not come in a tidy format and much of the data analysis work may involve making the dataset tidy. Once a dataset is tidy, it can be used as input into different functions that transform, model, or visualize this dataset.

## 10.2 THE "TIDYVERSE" PACKAGE

There are a number of R packages that take advantage of the tidy data form and can be used to do interesting things with data. The collection of packages is referred to as the "tidyverse" because of their dependence on tidy data.

"Tidyverse" packages include among others:

1. **dplyr:** a suite of functions for working with data frames.
2. **tidyr:** for tidying data.
3. **magrittr:** defines the %>% operator for chaining functions together in a series of operations on data.
4. **ggplot2:** for data visualization.

The "tidyverse" package can be used to install all of the packages in the tidyverse at once. For example, instead of starting an R Script with this:

```
library(dplyr)
library(tidyr)
library(ggplot2)
library(magrittr)
```

You can start with this:

```
library(tidyverse)
```

## 10.3 DPLYR PACKAGE

The data frame is a key data structure in statistics and in R. The basic structure of a data frame is that there is one observation per row and each column represents a variable, a measure, feature, or characteristic of that observation.

In previous chapters we have already discussed some tools like [ and $ operators to extract subsets of data frames. However, other operations, like filtering, re-ordering, and collapsing, can often be tedious operations in R whose syntax is not very intuitive. The dplyr package is designed to overcome a lot of these problems and to provide a highly optimized set of functions for dealing with data frames.

### 10.3.1 Key Functions

Some of the key "verbs" or functions provided by the dplyr package are

1. **Select:** return a subset of the columns of a data frame.
2. **Filter:** extract a subset of rows from a data frame based on logical conditions.
3. **Arrange:** reorder rows of a data frame
4. **Rename:** rename variables in a data frame
5. **Mutate:** add new variables/columns or transform existing variables
6. **Summarise/summarize:** generate summary statistics of different variables in the data frame, possibly within strata defined by group_by() function

All of the functions that we will discuss in this chapter will have a few common characteristics:

1. The first argument is a data frame.
2. The subsequent arguments describe what to do with the data frame specified in the first argument, and you can refer to columns in the data frame directly without using the $ operator (just use the column names).
3. The result of a function is a new data frame.
4. Data frames must be properly formatted for this to be useful. In particular, the data must be tidy as discussed before. In short, there should be one observation per row, and each column should represent a feature or characteristic of that observation.

### 10.3.2 Select

The select() function can be used to select columns of a data frame that you want to focus on. When you have a large data frame containing many columns and your analysis many only use a subset of columns, select() function allows you to get these few columns.

The regicor dataframe, from the package compareGroups and introduced in Chapter 6, contains data from three different cross-sectional surveys of individuals representative of the population from a north-west Spanish province (Girona), REGICOR study. Visit [www.regicor.org].

The data consists of 2294 observations (rows) and 25 variables (columns).

```
library(tidyverse)
## -- Attaching packages ---------------------------------------------------------
-------- tidyverse 1.3.0 --
## <U+2713> ggplot2 3.2.1  <U+2713> purrr 0.3.3
## <U+2713> tibble 2.1.3  <U+2713> dplyr 0.8.3
## <U+2713> tidyr 1.0.0  <U+2713> stringr 1.4.0
## <U+2713> readr 1.3.1  <U+2713> forcats 0.4.0
## -- Conflicts ----------------------------------------------------------------
tidyverse_conflicts() --
## x dplyr::filter() masks stats::filter()
## x dplyr::lag() masks stats::lag()
library(compareGroups)
## Warning: package 'compareGroups' was built under R version 3.6.3
## Loading required package: SNPassoc
## Warning: package 'SNPassoc' was built under R version 3.6.3
## Loading required package: haplo.stats
## Warning: package 'haplo.stats' was built under R version 3.6.3
## Loading required package: survival
## Warning: package 'survival' was built under R version 3.6.3
## Loading required package: mvtnorm
## Loading required package: parallel
## Registered S3 method overwritten by 'SNPassoc':
## method   from
## summary.haplo.glm haplo.stats
data(regicor)
```

```
str(regicor)
## 'data.frame': 2294 obs. of 25 variables:
## $ id  : num 2.26e+03 1.88e+03 3.00e+09 3.00e+09 3.00e+09 ...
## ..- attr(*, "label")= Named chr "Individual id"
## .. ..- attr(*, "names")= chr "id"
## $ year : Factor w/ 3 levels "1995","2000",..: 3 3 2 2 2 2 2 1 3 1 ...
## ..- attr(*, "label")= Named chr "Recruitment year"
## .. ..- attr(*, "names")= chr "year"
## $ age  : int 70 56 37 69 70 40 66 53 43 70 ...
## ..- attr(*, "label")= Named chr "Age"
## .. ..- attr(*, "names")= chr "age"
## $ sex  : Factor w/ 2 levels "Male","Female": 2 2 1 2 2 2 1 2 2 1 ...
## ..- attr(*, "label")= chr "Sex"
## $ smoker : Factor w/ 3 levels "Never smoker",..: 1 1 2 1 NA 2 1 1 3 3 ...
## ..- attr(*, "label")= Named chr "Smoking status"
## .. ..- attr(*, "names")= chr "smoker"
## $ sbp  : int 138 139 132 168 NA 108 120 132 95 142 ...
## ..- attr(*, "label")= Named chr "Systolic blood pressure"
## .. ..- attr(*, "names")= chr "sbp"
## $ dbp  : int 75 89 82 97 NA 70 72 78 65 78 ...
## ..- attr(*, "label")= Named chr "Diastolic blood pressure"
## .. ..- attr(*, "names")= chr "dbp"
## $ histhtn : Factor w/ 2 levels "Yes","No": 2 2 2 2 2 2 1 2 2 2 ...
## ..- attr(*, "label")= Named chr "History of hypertension"
## .. ..- attr(*, "names")= chr "histbp"
## $ txhtn : Factor w/ 2 levels "No","Yes": 1 1 1 1 1 1 2 1 1 1 ...
## ..- attr(*, "label")= chr "Hypertension treatment"
## $ chol : num 294 220 245 168 NA NA 298 254 194 188 ...
## ..- attr(*, "label")= Named chr "Total cholesterol"
## .. ..- attr(*, "names")= chr "chol"
```

```
## $ hdl  : num 57 50 59.8 53.2 NA ...
## ..- attr(*, "label")= Named chr "HDL cholesterol"
## .. ..- attr(*, "names")= chr "hdl"
## $ triglyc : num 93 160 89 116 NA 94 71 NA 68 137 ...
## ..- attr(*, "label")= Named chr "Triglycerides"
## .. ..- attr(*, "names")= chr "triglyc"
## $ ldl  : num 218.4 138 167.4 91.6 NA ...
## ..- attr(*, "label")= Named chr "LDL cholesterol"
## .. ..- attr(*, "names")= chr "ldl"
## $ histchol: Factor w/ 2 levels "Yes","No": 2 2 2 2 NA 2 1 2 2 2 ...
## ..- attr(*, "label")= chr "History of hyperchol."
## $ txchol : Factor w/ 2 levels "No","Yes": 1 1 1 1 NA 1 1 1 1 1 ...
## ..- attr(*, "label")= Named chr "Cholesterol treatment"
## .. ..- attr(*, "names")= chr "txchol"
## $ height : num 160 163 170 147 NA ...
## ..- attr(*, "label")= Named chr "Height (cm)"
## .. ..- attr(*, "names")= chr "height"
## $ weight : num 64 67 70 68 NA 43.5 79.2 45.8 53 62 ...
## ..- attr(*, "label")= Named chr "Weight (Kg)"
## .. ..- attr(*, "names")= chr "weight"
## $ bmi  : num 25 25.2 24.2 31.5 NA ...
## ..- attr(*, "label")= Named chr "Body mass index"
## .. ..- attr(*, "names")= chr "bmi"
## $ phyact : num 304 160 553 522 NA ...
## ..- attr(*, "label")= Named chr "Physical activity (Kcal/week)"
## .. ..- attr(*, "names")= chr "phyact"
## $ pcs  : num 54.5 58.2 43.4 54.3 NA ...
## ..- attr(*, "label")= Named chr "Physical component"
## .. ..- attr(*, "names")= chr "pcs"
## $ mcs  : num 58.9 48 62.6 57.9 NA ...
## ..- attr(*, "label")= chr "Mental component"
```

```
## $ cv : Factor w/ 2 levels "No","Yes": 1 1 1 1 NA 1 1 1 1 1 ...
## ..- attr(*, "label")= chr "Cardiovascular event"
## $ tocv : num 1025 2757 1906 1055 NA ...
## ..- attr(*, "label")= chr "Days to cardiovascular event or end of follow-up"
## $ death : Factor w/ 2 levels "No","Yes": 2 1 1 1 NA 1 2 1 1 1 ...
## ..- attr(*, "label")= chr "Overall death"
## $ todeath : num 1299.2 39.3 858.4 1833.1 NA ...
## ..- attr(*, "label")= chr "Days to overall death or end of follow-up"
head(regicor)
##    id year age sex     smoker sbp dbp histhtn txhtn
## 6101  2265 2005 70 Female   Never smoker 138 75  No No
## 5762  1882 2005 56 Female   Never smoker 139 89  No No
## 2992 3000105616 2000 37 Male Current or former < 1y 132 82  No No
## 2611 3000103485 2000 69 Female   Never smoker 168 97  No No
## 2762 3000103963 2000 70 Female     <NA> NA NA  No No
## 1516 3000100883 2000 40 Female Current or former < 1y 108 70  No No
##  chol  hdl triglyc  ldl histchol txchol height weight  bmi
## 6101 294 57.00000  93 218.40000  No  No 160 64.0 25.00000
## 5762 220 50.00000  160 138.00000  No  No 163 67.0 25.21736
## 2992 245 59.80429  89 167.39571  No  No 170 70.0 24.22145
## 2611 168 53.17571  116 91.62429  No  No 147 68.0 31.46837
## 2762 NA  NA  NA  NA  <NA> <NA>  NA  NA  NA
## 1516 NA 68.90000  94  NA  No  No 158 43.5 17.42509
##  phyact pcs mcs cv  tocv death todeath
## 6101 304.2000 54.455 58.918 No 1024.882 Yes 1299.16343
## 5762 160.3000 58.165 47.995 No 2756.849 No 39.32629
## 2992 552.7912 43.429 62.585 No 1905.969 No 858.42203
## 2611 522.0000 54.325 57.900 No 1055.380 No 1833.07619
## 2762  NA  NA  NA <NA>  NA <NA>   NA
## 1516 386.9505 57.315 47.869 No 3239.241 No 877.61155
```

To subset for only a single column, use its name or its number. For example, to select only the second column (year) into a new dataframe called dat.

```
dat<-select(regicor, year)

#or

dat2<-select(regicor,2)

identical(dat,dat2)
## [1] TRUE
str(dat)
## 'data.frame': 2294 obs. of 1 variable:
## $ year: Factor w/ 3 levels "1995","2000",..: 3 3 2 2 2 2 2 1 3 1 ...
## ..- attr(*, "label")= Named chr "Recruitment year"
## .. ..- attr(*, "names")= chr "year"
head(dat)
##  year
## 6101 2005
## 5762 2005
## 2992 2000
## 2611 2000
## 2762 2000
## 1516 2000
```

The subsetted dataframe contains one column (year) and 2294 rows.

To subset for multiple columns, use their name or their number. For example, to select only the second, third, and fourth column (year, age, sex) into a new dataframe called dat.

```
dat<-select(regicor, year,age,sex)
```

```
#or

dat2<-select(regicor,2:4)

identical(dat,dat2)
## [1] TRUE
str(dat)
## 'data.frame': 2294 obs. of 3 variables:
## $ year: Factor w/ 3 levels "1995","2000",..: 3 3 2 2 2 2 2 1 3 1 ...
## ..- attr(*, "label")= Named chr "Recruitment year"
## .. ..- attr(*, "names")= chr "year"
## $ age : int 70 56 37 69 70 40 66 53 43 70 ...
## ..- attr(*, "label")= Named chr "Age"
## .. ..- attr(*, "names")= chr "age"
## $ sex : Factor w/ 2 levels "Male","Female": 2 2 1 2 2 2 1 2 2 1 ...
## ..- attr(*, "label")= chr "Sex"
head(dat)
##  year age sex
## 6101 2005 70 Female
## 5762 2005 56 Female
## 2992 2000 37 Male
## 2611 2000 69 Female
## 2762 2000 70 Female
## 1516 2000 40 Female
```

The subsetted dataframe contains 3 columns (year,age,sex) and 2294 rows. To subset for all columns except the second, third, and fourth column (year, age, sex), use the - sign.

```
dat<-select(regicor, -year,-age,-sex)
```

```
#or

dat2<-select(regicor,-2:-4)

identical(dat,dat2)
## [1] TRUE
str(dat)
## 'data.frame': 2294 obs. of 22 variables:
## $ id  : num 2.26e+03 1.88e+03 3.00e+09 3.00e+09 3.00e+09 ...
## ..- attr(*, "label")= Named chr "Individual id"
## .. ..- attr(*, "names")= chr "id"
## $ smoker : Factor w/ 3 levels "Never smoker",..: 1 1 2 1 NA 2 1 1 3 3 ...
## ..- attr(*, "label")= Named chr "Smoking status"
## .. ..- attr(*, "names")= chr "smoker"
## $ sbp  : int 138 139 132 168 NA 108 120 132 95 142 ...
## ..- attr(*, "label")= Named chr "Systolic blood pressure"
## .. ..- attr(*, "names")= chr "sbp"
## $ dbp  : int 75 89 82 97 NA 70 72 78 65 78 ...
## ..- attr(*, "label")= Named chr "Diastolic blood pressure"
## .. ..- attr(*, "names")= chr "dbp"
## $ histhtn : Factor w/ 2 levels "Yes","No": 2 2 2 2 2 2 1 2 2 2 ...
## ..- attr(*, "label")= Named chr "History of hypertension"
## .. ..- attr(*, "names")= chr "histbp"
## $ txhtn : Factor w/ 2 levels "No","Yes": 1 1 1 1 1 1 2 1 1 1 ...
## ..- attr(*, "label")= chr "Hypertension treatment"
## $ chol : num 294 220 245 168 NA NA 298 254 194 188 ...
## ..- attr(*, "label")= Named chr "Total cholesterol"
## .. ..- attr(*, "names")= chr "chol"
## $ hdl  : num 57 50 59.8 53.2 NA ...
## ..- attr(*, "label")= Named chr "HDL cholesterol"
```

```
## .. ..- attr(*, "names")= chr "hdl"
## $ triglyc : num 93 160 89 116 NA 94 71 NA 68 137 ...
## ..- attr(*, "label")= Named chr "Triglycerides"
## .. ..- attr(*, "names")= chr "triglyc"
## $ ldl  : num 218.4 138 167.4 91.6 NA ...
## ..- attr(*, "label")= Named chr "LDL cholesterol"
## .. ..- attr(*, "names")= chr "ldl"
## $ histchol: Factor w/ 2 levels "Yes","No": 2 2 2 2 NA 2 1 2 2 2 ...
## ..- attr(*, "label")= chr "History of hyperchol."
## $ txchol : Factor w/ 2 levels "No","Yes": 1 1 1 1 NA 1 1 1 1 1 ...
## ..- attr(*, "label")= Named chr "Cholesterol treatment"
## .. ..- attr(*, "names")= chr "txchol"
## $ height : num 160 163 170 147 NA ...
## ..- attr(*, "label")= Named chr "Height (cm)"
## .. ..- attr(*, "names")= chr "height"
## $ weight : num 64 67 70 68 NA 43.5 79.2 45.8 53 62 ...
## ..- attr(*, "label")= Named chr "Weight (Kg)"
## .. ..- attr(*, "names")= chr "weight"
## $ bmi  : num 25 25.2 24.2 31.5 NA ...
## ..- attr(*, "label")= Named chr "Body mass index"
## .. ..- attr(*, "names")= chr "bmi"
## $ phyact : num 304 160 553 522 NA ...
## ..- attr(*, "label")= Named chr "Physical activity (Kcal/week)"
## .. ..- attr(*, "names")= chr "phyact"
## $ pcs  : num 54.5 58.2 43.4 54.3 NA ...
## ..- attr(*, "label")= Named chr "Physical component"
## .. ..- attr(*, "names")= chr "pcs"
## $ mcs  : num 58.9 48 62.6 57.9 NA ...
## ..- attr(*, "label")= chr "Mental component"
## $ cv  : Factor w/ 2 levels "No","Yes": 1 1 1 1 NA 1 1 1 1 1 ...
## ..- attr(*, "label")= chr "Cardiovascular event"
```

```
## $ tocv : num 1025 2757 1906 1055 NA ...
## ..- attr(*, "label")= chr "Days to cardiovascular event or end of follow-up"
## $ death : Factor w/ 2 levels "No","Yes": 2 1 1 1 NA 1 2 1 1 1 ...
## ..- attr(*, "label")= chr "Overall death"
## $ todeath : num 1299.2 39.3 858.4 1833.1 NA ...
## ..- attr(*, "label")= chr "Days to overall death or end of follow-up"
head(dat)
##    id    smoker sbp dbp histhtn txhtn chol  hdl
## 6101  2265   Never smoker 138 75  No No 294 57.00000
## 5762  1882   Never smoker 139 89  No No 220 50.00000
## 2992 3000105616 Current or former < 1y 132 82  No No 245 59.80429
## 2611 3000103485   Never smoker 168 97  No No 168 53.17571
## 2762 3000103963     <NA> NA NA  No No NA  NA
## 1516 3000100883 Current or former < 1y 108 70  No No NA 68.90000
##  triglyc  ldl histchol txchol height weight  bmi phyact pcs
## 6101  93 218.40000  No  No 160 64.0 25.00000 304.2000 54.455
## 5762  160 138.00000  No  No 163 67.0 25.21736 160.3000 58.165
## 2992  89 167.39571  No  No 170 70.0 24.22145 552.7912 43.429
## 2611  116 91.62429  No  No 147 68.0 31.46837 522.0000 54.325
## 2762  NA  NA  <NA> <NA>  NA  NA  NA  NA  NA
## 1516  94  NA  No  No 158 43.5 17.42509 386.9505 57.315
##   mcs cv  tocv death todeath
## 6101 58.918 No 1024.882 Yes 1299.16343
## 5762 47.995 No 2756.849 No 39.32629
## 2992 62.585 No 1905.969 No 858.42203
## 2611 57.900 No 1055.380 No 1833.07619
## 2762  NA <NA>  NA <NA>   NA
## 1516 47.869 No 3239.241 No 877.61155
```

The subsetted dataframe contains 22 columns (25–3) and 2294 rows.

Other useful arguments of the select() function that allow subsetting for multiple columns:

starts_with: select all columns that start with a certain string

ends_with: select all columns that end with a certain string

contains: select all columns that include a certain string

Example of starts_with argument to select all columns that begins with "h".

```
dat<-select(regicor, starts_with("h"))

str(dat)
## 'data.frame': 2294 obs. of 4 variables:
## $ histhtn : Factor w/ 2 levels "Yes","No": 2 2 2 2 2 2 1 2 2 2 ...
## ..- attr(*, "label")= Named chr "History of hypertension"
## .. ..- attr(*, "names")= chr "histbp"
## $ hdl  : num 57 50 59.8 53.2 NA ...
## ..- attr(*, "label")= Named chr "HDL cholesterol"
## .. ..- attr(*, "names")= chr "hdl"
## $ histchol: Factor w/ 2 levels "Yes","No": 2 2 2 2 NA 2 1 2 2 2 ...
## ..- attr(*, "label")= chr "History of hyperchol."
## $ height : num 160 163 170 147 NA ...
## ..- attr(*, "label")= Named chr "Height (cm)"
## .. ..- attr(*, "names")= chr "height"
head(dat)
##  histhtn  hdl histchol height
## 6101  No 57.00000  No 160
## 5762  No 50.00000  No 163
## 2992  No 59.80429  No 170
## 2611  No 53.17571  No 147
## 2762  No  NA  <NA>  NA
## 1516  No 68.90000  No 158
```

The subsetted dataframe contains 4 columns ("histhtn", "hdl", "histchol", "height") and 2294 rows. All columns start with "h".

Example of ends_with argument to select all columns that ends with "h".

```
dat<-select(regicor, ends_with("h"))

str(dat)
## 'data.frame': 2294 obs. of 2 variables:
## $ death : Factor w/ 2 levels "No","Yes": 2 1 1 1 NA 1 2 1 1 1 ...
## ..- attr(*, "label")= chr "Overall death"
## $ todeath: num 1299.2 39.3 858.4 1833.1 NA ...
## ..- attr(*, "label")= chr "Days to overall death or end of follow-up"
head(dat)
##  death todeath
## 6101 Yes 1299.16343
## 5762 No 39.32629
## 2992 No 858.42203
## 2611 No 1833.07619
## 2762 <NA>   NA
## 1516 No 877.61155
```

The subsetted dataframe contains 2 columns ("death", "todeath") and 2294 rows. All columns end with "h".

Example of contains argument to select all columns that contain "hist".

```
dat<-select(regicor, contains("hist"))

str(dat)
## 'data.frame': 2294 obs. of 2 variables:
## $ histhtn : Factor w/ 2 levels "Yes","No": 2 2 2 2 2 2 1 2 2 2 ...
## ..- attr(*, "label")= Named chr "History of hypertension"
## .. ..- attr(*, "names")= chr "histbp"
```

```
## $ histchol: Factor w/ 2 levels "Yes","No": 2 2 2 2 NA 2 1 2 2 2 ...
## ..- attr(*, "label")= chr "History of hyperchol."
head(dat)
## histhtn histchol
## 6101 No No
## 5762 No No
## 2992 No No
## 2611 No No
## 2762 No <NA>
## 1516 No No
```

The subsetted dataframe contains 2 columns ("histhtn", "histchol") and 2294 rows. All columns contain "hist".

### 10.3.3 Filter

While select picks out certain columns of the data frame, filter picks out certain rows. With filter, you can specify certain conditions using R's logical operators, and the function will return rows that meet those conditions.

R's logical operators include:

*== Equals,*

*!= Does not equal,*

*> Greater than,*

*>= Greater than or equal to,*

*< Less than*

*<= Less than or equal to*

*%in% Included in*

*is.na() Is a missing value*

If you are unsure of how to write a logical statement, but know how to write its opposite, you can use the ! operator to negate the whole statement. A common use of this is to identify observations with non-missing data (e.g., !(is.na(column names))).

When you use a logical statement within filter for a certain column, it will return just the rows of the dataframe where the logical statement is true.

If you would like to use several logical conditions together and select rows where all or any of the conditions are true, you can use the "and" (&) or "or" (|) operators.

Example, subset the regicor dataframe to include only rows with bmi >30 in a dataframe called dat.

```
dat<-filter(regicor, bmi >30)
```

```
str(dat)
## 'data.frame': 637 obs. of 25 variables:
## $ id  : num 3.00e+09 2.07e+03 3.00e+09 3.00e+09 2.86e+03 ...
## ..- attr(*, "label")= Named chr "Individual id"
## .. ..- attr(*, "names")= chr "id"
## $ year : Factor w/ 3 levels "1995","2000",..: 2 3 2 2 3 3 3 2 2 2 ...
## $ age  : int 69 70 70 35 60 71 65 73 70 58 ...
## ..- attr(*, "label")= Named chr "Age"
## .. ..- attr(*, "names")= chr "age"
## $ sex  : Factor w/ 2 levels "Male","Female": 2 1 1 2 2 2 2 1 2 2 ...
## $ smoker : Factor w/ 3 levels "Never smoker",..: 1 3 3 1 1 1 1 3 1 1 ...
## $ sbp  : int 168 195 160 110 119 132 147 140 142 150 ...
## ..- attr(*, "label")= Named chr "Systolic blood pressure"
## .. ..- attr(*, "names")= chr "sbp"
## $ dbp  : int 97 97 86 70 55 81 82 80 74 90 ...
## ..- attr(*, "label")= Named chr "Diastolic blood pressure"
## .. ..- attr(*, "names")= chr "dbp"
## $ histhtn : Factor w/ 2 levels "Yes","No": 2 1 2 2 2 1 1 1 2 1 ...
## $ txhtn : Factor w/ 2 levels "No","Yes": 1 2 1 1 1 2 2 2 1 2 ...
## $ chol : num 168 182 230 197 244 163 212 239 258 202 ...
## ..- attr(*, "label")= Named chr "Total cholesterol"
## .. ..- attr(*, "names")= chr "chol"
## $ hdl  : num 53.2 34 45.7 44.5 51 ...
## ..- attr(*, "label")= Named chr "HDL cholesterol"
```

```
## .. ..- attr(*, "names")= chr "hdl"
## $ triglyc : num 116 113 92 73 114 110 92 156 93 103 ...
## ..- attr(*, "label")= Named chr "Triglycerides"
## .. ..- attr(*, "names")= chr "triglyc"
## $ ldl  : num 91.6 125.4 165.9 137.9 170.2 ...
## ..- attr(*, "label")= Named chr "LDL cholesterol"
## .. ..- attr(*, "names")= chr "ldl"
## $ histchol: Factor w/ 2 levels "Yes","No": 2 2 1 2 2 2 2 1 1 2 ...
## $ txchol : Factor w/ 2 levels "No","Yes": 1 1 2 1 1 1 1 2 1 1 ...
## $ height : num 147 156 170 164 158 ...
## ..- attr(*, "label")= Named chr "Height (cm)"
## .. ..- attr(*, "names")= chr "height"
## $ weight : num 68 80 95.5 84.5 94 68 98 89 77 94 ...
## ..- attr(*, "label")= Named chr "Weight (Kg)"
## .. ..- attr(*, "names")= chr "weight"
## $ bmi  : num 31.5 32.9 33 31.2 37.7 ...
## ..- attr(*, "label")= Named chr "Body mass index"
## .. ..- attr(*, "names")= chr "bmi"
## $ phyact : num 522 149 360 451 28 ...
## ..- attr(*, "label")= Named chr "Physical activity (Kcal/week)"
## .. ..- attr(*, "names")= chr "phyact"
## $ pcs  : num 54.3 NA 40.7 51.3 40.7 ...
## ..- attr(*, "label")= Named chr "Physical component"
## .. ..- attr(*, "names")= chr "pcs"
## $ mcs  : num 57.9 NA 55.9 57.6 57.6 ...
## ..- attr(*, "label")= chr "Mental component"
## $ cv  : Factor w/ 2 levels "No","Yes": 1 1 2 1 1 2 1 1 1 1 ...
## $ tocv : num 1055.4 774.3 10.1 776.8 1105.5 ...
## ..- attr(*, "label")= chr "Days to cardiovascular event or end of follow-up"
## $ death : Factor w/ 2 levels "No","Yes": 1 1 1 NA 1 1 1 1 1 1 ...
```

```
## $ todeath : num 1833 934 434 NA 1523 ...
## ..- attr(*, "label")= chr "Days to overall death or end of follow-up"
summary(dat$bmi)
## Min. 1st Qu. Median Mean 3rd Qu. Max.
## 30.00 30.90 32.30 33.27 34.60 48.24
head(dat)
##   id year age sex  smoker sbp dbp histhtn txhtn chol  hdl
## 1 3000103485 2000 69 Female Never smoker 168 97  No No 168 53.17571
## 2  2071 2005 70 Male Former >= 1y 195 97  Yes Yes 182 34.00000
## 3 3000108163 2000 70 Male Former >= 1y 160 86  No No 230 45.74714
## 4 3000101478 2000 35 Female Never smoker 110 70  No No 197 44.49000
## 5  2856 2005 60 Female Never smoker 119 55  No No 244 51.00000
## 6  562 2005 71 Female Never smoker 132 81  Yes Yes 163 61.00000
## triglyc  ldl histchol txchol height weight  bmi phyact pcs
## 1  116 91.62429  No  No 147.0 68.0 31.46837 522.0000 54.325
## 2  113 125.40000  No  No 156.0 80.0 32.87311 148.6000  NA
## 3  92 165.85286  Yes Yes 170.0 95.5 33.04498 359.7132 40.736
## 4  73 137.91000  No  No 164.5 84.5 31.22661 450.9890 51.314
## 5  114 170.20000  No  No 158.0 94.0 37.65422 28.0000 40.744
## 6  110 80.00000  No  No 149.0 68.0 30.62925 386.5000 30.381
##  mcs cv  tocv death todeath
## 1 57.900 No 1055.37962 No 1833.0762
## 2  NA No 774.31624 No 934.1619
## 3 55.914 Yes 10.10702 No 434.0974
## 4 57.612 No 776.81785 <NA>  NA
## 5 57.586 No 1105.47313 No 1523.4886
## 6 47.051 Yes 1215.52144 No 1155.6976
```

The filtered dataframe now consists of only 637 rows and using the summary on the bmi column of the new dataframe confirmed that minimum value is

30 (in fact, it is 30.004 but reported 30.00 by rounding).

Example, subset the regicor dataframe to include only rows with bmi >30 and heights > 180 in a dataframe called dat.

```
dat<-filter(regicor, bmi >30 & height > 180)
```

```
str(dat)
## 'data.frame': 10 obs. of 25 variables:
## $ id  : num 3.00e+09 3.67e+03 3.00e+09 3.00e+09 3.00e+09 ...
## ..- attr(*, "label")= Named chr "Individual id"
## .. ..- attr(*, "names")= chr "id"
## $ year : Factor w/ 3 levels "1995","2000",..: 2 3 2 2 2 3 1 3 3 3
## $ age  : int 38 39 40 44 44 66 58 54 36 39
## ..- attr(*, "label")= Named chr "Age"
## .. ..- attr(*, "names")= chr "age"
## $ sex  : Factor w/ 2 levels "Male","Female": 1 1 1 1 1 1 1 1 1 1
## $ smoker : Factor w/ 3 levels "Never smoker",..: 2 2 2 3 3 2 3 2 3 2
## $ sbp  : int 130 118 124 150 150 125 132 145 118 120
## ..- attr(*, "label")= Named chr "Systolic blood pressure"
## .. ..- attr(*, "names")= chr "sbp"
## $ dbp  : int 92 80 85 84 80 89 84 88 76 71
## ..- attr(*, "label")= Named chr "Diastolic blood pressure"
## .. ..- attr(*, "names")= chr "dbp"
## $ histhtn : Factor w/ 2 levels "Yes","No": 2 2 1 1 1 1 2 2 2 NA
## $ txhtn : Factor w/ 2 levels "No","Yes": 1 1 1 1 1 2 1 1 1 1
## $ chol : num 161 210 176 241 NA 205 216 259 239 151
## ..- attr(*, "label")= Named chr "Total cholesterol"
## .. ..- attr(*, "names")= chr "chol"
## $ hdl  : num 49.9 35 58.3 38.3 39.5 ...
## ..- attr(*, "label")= Named chr "HDL cholesterol"
## .. ..- attr(*, "names")= chr "hdl"
```

```
## $ triglyc : num 58 120 113 101 168 129 84 215 102 77
## ..- attr(*, "label")= Named chr "Triglycerides"
## .. ..- attr(*, "names")= chr "triglyc"
## $ ldl  : num 99.5 151 95.1 182.5 NA ...
## ..- attr(*, "label")= Named chr "LDL cholesterol"
## .. ..- attr(*, "names")= chr "ldl"
## $ histchol: Factor w/ 2 levels "Yes","No": 2 2 2 1 1 2 2 1 1 2
## $ txchol : Factor w/ 2 levels "No","Yes": 1 1 1 1 1 1 1 2 1 1
## $ height : num 185 182 182 189 182 ...
## ..- attr(*, "label")= Named chr "Height (cm)"
## .. ..- attr(*, "names")= chr "height"
## $ weight : num 117 121 104 110 107 ...
## ..- attr(*, "label")= Named chr "Weight (Kg)"
## .. ..- attr(*, "names")= chr "weight"
## $ bmi  : num 34.2 36.5 31.2 30.9 32.3 ...
## ..- attr(*, "label")= Named chr "Body mass index"
## .. ..- attr(*, "names")= chr "bmi"
## $ phyact : num 236 201 266 423 354 ...
## ..- attr(*, "label")= Named chr "Physical activity (Kcal/week)"
## .. ..- attr(*, "names")= chr "phyact"
## $ pcs  : num 50.8 49.8 56.9 60.7 NA ...
## ..- attr(*, "label")= Named chr "Physical component"
## .. ..- attr(*, "names")= chr "pcs"
## $ mcs  : num 35.3 59.2 50.6 20.5 NA ...
## ..- attr(*, "label")= chr "Mental component"
## $ cv  : Factor w/ 2 levels "No","Yes": 1 1 1 NA NA 1 1 1 1 1
## $ tocv : num 3105 2227 1523 NA NA ...
## ..- attr(*, "label")= chr "Days to cardiovascular event or end of follow-up"
## $ death : Factor w/ 2 levels "No","Yes": 1 1 1 1 1 1 1 2 1 1
## $ todeath : num 1492 1278 3582 1429 3252 ...
```

```
## ..- attr(*, "label")= chr "Days to overall death or end of follow-up"
summary(dat$bmi)
## Min. 1st Qu. Median Mean 3rd Qu. Max.
## 30.83 31.13 32.28 32.81 33.87 36.53
summary(dat$height)
## Min. 1st Qu. Median Mean 3rd Qu. Max.
## 181.0 182.0 182.0 183.2 184.1 189.0
head(dat)
##   id year age sex    smoker sbp dbp histhtn txhtn chol
## 1 3000105416 2000 38 Male Current or former < 1y 130 92  No No 161
## 2  3671 2005 39 Male Current or former < 1y 118 80  No No 210
## 3 3000105544 2000 40 Male Current or former < 1y 124 85  Yes No 176
## 4 3000105325 2000 44 Male   Former >= 1y 150 84  Yes No 241
## 5 3000106065 2000 44 Male   Former >= 1y 150 80  Yes No NA
## 6  2994 2005 66 Male Current or former < 1y 125 89  Yes Yes 205
##  hdl triglyc  ldl histchol txchol height weight  bmi phyact
## 1 49.86143  58 99.53857  No  No 185 117.0 34.18554 235.5907
## 2 35.00000  120 151.00000  No  No 182 121.0 36.52940 200.6000
## 3 58.31857  113 95.08143  No  No 182 103.5 31.24623 265.9615
## 4 38.31857  101 182.48143  Yes  No 189 110.5 30.93418 422.6374
## 5 39.50000  168  NA  Yes  No 182 107.0 32.30286 353.6538
## 6 40.00000  129 139.20000  No  No 183 108.0 32.24940 566.0000
##  pcs mcs cv  tocv death todeath
## 1 50.797 35.302 No 3104.723 No 1492.0831
## 2 49.776 59.179 No 2226.968 No 1277.9946
## 3 56.947 50.580 No 1523.464 No 3581.7350
## 4 60.739 20.481 <NA>  NA No 1429.4829
## 5  NA  NA <NA>  NA No 3252.2169
## 6 48.153 59.084 No 3243.830 No 729.6281
```

The filtered dataframe now consists of only 10 rows (only 10 observations or 10 participants stratify these criteria) and using the summary on the bmi and height columns of the new dataframe confirmed that minimum value is 30 for bmi and 180 for height.

Example from storms dataset

The storms is dataframe in the dplyr package and is a subset of the NOAA Atlantic hurricane database best track data, http://www.nhc.noaa.gov/data/#hurdat. The data includes the positions and attributes of 198 tropical storms, measured every six hours during the lifetime of a storm.

The data consists of 10,010 observations (rows) and 13 variables (columns):

1. name: Storm Name

2, 3, 4. year,month,day: Date of report

5. hour: Hour of report (in UTC)

6, 7. lat,long: Location of storm center

8. status: Storm classification (Tropical Depression, Tropical Storm, or Hurricane)

9. category: Saffir-Simpson storm category (estimated from wind speed. –1 = Tropical Depression, 0 = Tropical Storm)

10. wind: storm's maximum sustained wind speed (in knots)

11. pressure: Air pressure at the storm's center (in millibars)

12. ts_diameter: Diameter of the area experiencing tropical storm strength winds (34 knots or above)

13. hu_diameter: Diameter of the area experiencing hurricane strength winds (64 knots or above)

To filter for rows of Katrina and Keith storms in a new dataframe called dat

```
data("storms")

str(storms)
## Classes 'tbl_df', 'tbl' and 'data.frame': 10010 obs. of 13 variables:
## $ name  : chr "Amy" "Amy" "Amy" "Amy" ...
## $ year  : num 1975 1975 1975 1975 1975 ...
## $ month  : num 6 6 6 6 6 6 6 6 6 6 ...
```

```
## $ day  : int 27 27 27 27 28 28 28 28 29 29 ...
## $ hour  : num 0 6 12 18 0 6 12 18 0 6 ...
## $ lat  : num 27.5 28.5 29.5 30.5 31.5 32.4 33.3 34 34.4 34 ...
## $ long  : num –79 –79 –79 –79 –78.8 –78.7 –78 –77 –75.8 –74.8 ...
## $ status  : chr “tropical depression” “tropical depression” “tropical depression” “tropical depression” ...
## $ category : Ord.factor w/ 7 levels “–1”<“0”<“1”<“2”<..: 1 1 1 1 1 1 1 1 2 2 ...
## $ wind  : int 25 25 25 25 25 25 25 30 35 40 ...
## $ pressure : int 1013 1013 1013 1013 1012 1012 1011 1006 1004 1002 ...
## $ ts_diameter: num NA NA NA NA NA NA NA NA NA NA ...
## $ hu_diameter: num NA NA NA NA NA NA NA NA NA NA ...
head(storms)
## # A tibble: 6 x 13
## name year month day hour lat long status category wind pressure
## <chr> <dbl> <dbl> <int> <dbl> <dbl> <dbl> <chr> <ord> <int> <int>
## 1 Amy 1975  6 27  0 27.5 –79 tropi… –1   25  1013
## 2 Amy 1975  6 27  6 28.5 –79 tropi… –1   25  1013
## 3 Amy 1975  6 27 12 29.5 –79 tropi… –1   25  1013
## 4 Amy 1975  6 27 18 30.5 –79 tropi… –1   25  1013
## 5 Amy 1975  6 28  0 31.5 –78.8 tropi… –1   25  1012
## 6 Amy 1975  6 28  6 32.4 –78.7 tropi… –1   25  1012
## # … with 2 more variables: ts_diameter <dbl>, hu_diameter <dbl>
dat<-filter(storms, name %in% c(“Katrina”, “Keith”))

str(dat)
## Classes ‘tbl_df’, ‘tbl’ and ‘data.frame’: 133 obs. of 13 variables:
## $ name  : chr “Katrina” “Katrina” “Katrina” “Katrina” ...
## $ year  : num 1981 1981 1981 1981 1981 ...
## $ month  : num 11 11 11 11 11 11 11 11 11 11 ...
## $ day  : int 3 3 3 3 4 4 4 4 5 5 ...
## $ hour  : num 0 6 12 18 0 6 12 18 0 6 ...
```

```
## $ lat : num 16.9 17.2 17.5 17.8 18.1 18.3 18.6 18.9 19.2 19.6 ...
## $ long : num –81.2 –81.3 –81.4 –81.4 –81.4 –81.4 –81.3 –81.2 –81.1 –80.8 ...
## $ status : chr "tropical depression" "tropical depression" "tropical depression" "tropical depression" ...
## $ category : Ord.factor w/ 7 levels "–1"<"0"<"1"<"2"<..: 1 1 1 1 1 2 2 2 2 3 ...
## $ wind : int 25 25 25 30 30 35 40 50 60 65 ...
## $ pressure : int 1005 1005 1004 1002 1001 1000 998 996 993 988 ...
## $ ts_diameter: num NA NA NA NA NA NA NA NA NA NA ...
## $ hu_diameter: num NA NA NA NA NA NA NA NA NA NA ...
head(dat)
## # A tibble: 6 x 13
## name year month day hour lat long status category wind pressure
## <chr> <dbl> <dbl> <int> <dbl> <dbl> <dbl> <chr> <ord> <int> <int>
## 1 Katr… 1981 11 3 0 16.9 –81.2 tropi… –1 25 1005
## 2 Katr… 1981 11 3 6 17.2 –81.3 tropi… –1 25 1005
## 3 Katr… 1981 11 3 12 17.5 –81.4 tropi… –1 25 1004
## 4 Katr… 1981 11 3 18 17.8 –81.4 tropi… –1 30 1002
## 5 Katr… 1981 11 4 0 18.1 –81.4 tropi… –1 30 1001
## 6 Katr… 1981 11 4 6 18.3 –81.4 tropi… 0 35 1000
## # … with 2 more variables: ts_diameter <dbl>, hu_diameter <dbl>
```

The subsetted data contains only 133 rows for only Katrina and Keith storms. To filter for rows of Katrina and Keith storms and non-missing values in ts_diameter and hu_diameter in a new dataframe called dat.

```
data("storms")

str(storms)
## Classes 'tbl_df', 'tbl' and 'data.frame': 10010 obs. of 13 variables:
## $ name : chr "Amy" "Amy" "Amy" "Amy" ...
```

```
## $ year  : num 1975 1975 1975 1975 1975 ...
## $ month  : num 6 6 6 6 6 6 6 6 6 6 ...
## $ day  : int 27 27 27 27 28 28 28 28 29 29 ...
## $ hour  : num 0 6 12 18 0 6 12 18 0 6 ...
## $ lat  : num 27.5 28.5 29.5 30.5 31.5 32.4 33.3 34 34.4 34 ...
## $ long  : num –79 –79 –79 –79 –78.8 –78.7 –78 –77 –75.8 –74.8 ...
## $ status  : chr "tropical depression" "tropical depression" "tropical depression" "tropical depression" ...
## $ category : Ord.factor w/ 7 levels "–1"<"0"<"1"<"2"<..: 1 1 1 1 1 1 1 1 2 2 ...
## $ wind  : int 25 25 25 25 25 25 25 30 35 40 ...
## $ pressure : int 1013 1013 1013 1013 1012 1012 1011 1006 1004 1002 ...
## $ ts_diameter: num NA NA NA NA NA NA NA NA NA NA ...
## $ hu_diameter: num NA NA NA NA NA NA NA NA NA NA ...
head(storms)
## # A tibble: 6 x 13
## name year month day hour lat long status category wind pressure
## <chr> <dbl> <dbl> <int> <dbl> <dbl> <dbl> <chr> <ord> <int> <int>
## 1 Amy 1975  6 27  0 27.5 –79 tropi… –1   25  1013
## 2 Amy 1975  6 27  6 28.5 –79 tropi… –1   25  1013
## 3 Amy 1975  6 27 12 29.5 –79 tropi… –1   25  1013
## 4 Amy 1975  6 27 18 30.5 –79 tropi… –1   25  1013
## 5 Amy 1975  6 28  0 31.5 –78.8 tropi… –1   25  1012
## 6 Amy 1975  6 28  6 32.4 –78.7 tropi… –1   25  1012
## # … with 2 more variables: ts_diameter <dbl>, hu_diameter <dbl>
dat<-filter(storms, name %in% c("Katrina", "Keith") & !is.na(ts_diameter) &

   !is.na(hu_diameter))


str(dat)
## Classes 'tbl_df', 'tbl' and 'data.frame': 29 obs. of 13 variables:
```

```
## $ name : chr "Katrina" "Katrina" "Katrina" "Katrina" ...
## $ year : num 2005 2005 2005 2005 2005 ...
## $ month : num 8 8 8 8 8 8 8 8 8 8 ...
## $ day : int 23 24 24 24 24 25 25 25 25 26 ...
## $ hour : num 18 0 6 12 18 0 6 12 18 0 ...
## $ lat : num 23.1 23.4 23.8 24.5 25.4 26 26.1 26.2 26.2 25.9 ...
## $ long : num –75.1 –75.7 –76.2 –76.5 –76.9 –77.7 –78.4 –79 –79.6 –80.3 ...
## $ status : chr "tropical depression" "tropical depression" "tropical depression" "tropical storm" ...
## $ category : Ord.factor w/ 7 levels "–1"<"0"<"1"<"2"<..: 1 1 1 2 2 2 2 2 2 3 ...
## $ wind : int 30 30 30 35 40 45 50 55 60 70 ...
## $ pressure : int 1008 1007 1007 1006 1003 1000 997 994 988 983 ...
## $ ts_diameter: num 0 0 0 69 69 ...
## $ hu_diameter: num 0 0 0 0 0 ...
head(dat)
## # A tibble: 6 x 13
## name year month day hour lat long status category wind pressure
## <chr> <dbl> <dbl> <int> <dbl> <dbl> <dbl> <chr> <ord> <int> <int>
## 1 Katr… 2005 8 23 18 23.1 –75.1 tropi… –1 30 1008
## 2 Katr… 2005 8 24 0 23.4 –75.7 tropi… –1 30 1007
## 3 Katr… 2005 8 24 6 23.8 –76.2 tropi… –1 30 1007
## 4 Katr… 2005 8 24 12 24.5 –76.5 tropi… 0 35 1006
## 5 Katr… 2005 8 24 18 25.4 –76.9 tropi… 0 40 1003
## 6 Katr… 2005 8 25 0 26 –77.7 tropi… 0 45 1000
## # … with 2 more variables: ts_diameter <dbl>, hu_diameter <dbl>
```

The subsetted data contains only 29 rows for only Katrina and Keith storms and no missing data in ts_diameter and hu_diameter columns.

To filter for all storms except Katrina and Keith storms and non-missing values in ts_diameter and hu_diameter in a new dataframe called dat.

```
data(“storms”)

str(storms)
## Classes ‘tbl_df’, ‘tbl’ and ‘data.frame’: 10010 obs. of 13 variables:
## $ name : chr “Amy” “Amy” “Amy” “Amy” ...
## $ year : num 1975 1975 1975 1975 1975 ...
## $ month : num 6 6 6 6 6 6 6 6 6 6 ...
## $ day : int 27 27 27 27 28 28 28 28 29 29 ...
## $ hour : num 0 6 12 18 0 6 12 18 0 6 ...
## $ lat : num 27.5 28.5 29.5 30.5 31.5 32.4 33.3 34 34.4 34 ...
## $ long : num –79 –79 –79 –79 –78.8 –78.7 –78 –77 –75.8 –74.8 ...
## $ status : chr “tropical depression” “tropical depression” “tropical depression” “tropical depression” ...
## $ category : Ord.factor w/ 7 levels “–1”<“0”<“1”<“2”<..: 1 1 1 1 1 1 1 2 2 ...
## $ wind : int 25 25 25 25 25 25 25 30 35 40 ...
## $ pressure : int 1013 1013 1013 1013 1012 1012 1011 1006 1004 1002 ...
## $ ts_diameter: num NA NA NA NA NA NA NA NA NA NA ...
## $ hu_diameter: num NA NA NA NA NA NA NA NA NA NA ...
head(storms)
## # A tibble: 6 x 13
## name year month day hour lat long status category wind pressure
## <chr> <dbl> <dbl> <int> <dbl> <dbl> <dbl> <chr> <ord> <int> <int>
## 1 Amy 1975 6 27 0 27.5 –79 tropi… –1 25 1013
## 2 Amy 1975 6 27 6 28.5 –79 tropi… –1 25 1013
## 3 Amy 1975 6 27 12 29.5 –79 tropi… –1 25 1013
## 4 Amy 1975 6 27 18 30.5 –79 tropi… –1 25 1013
## 5 Amy 1975 6 28 0 31.5 –78.8 tropi… –1 25 1012
## 6 Amy 1975 6 28 6 32.4 –78.7 tropi… –1 25 1012
## # … with 2 more variables: ts_diameter <dbl>, hu_diameter <dbl>
dat<-filter(storms, !name %in% c(“Katrina”, “Keith”) & !is.na(ts_
```

```
diameter) &

  !is.na(hu_diameter))

str(dat)
## Classes 'tbl_df', 'tbl' and 'data.frame': 3453 obs. of 13 variables:
## $ name  : chr "Alex" "Alex" "Alex" "Alex" ...
## $ year  : num 2004 2004 2004 2004 2004 ...
## $ month  : num 7 8 8 8 8 8 8 8 8 8 ...
## $ day  : int 31 1 1 1 1 2 2 2 2 3 ...
## $ hour  : num 18 0 6 12 18 0 6 12 18 0 ...
## $ lat  : num 30.3 31 31.5 31.6 31.6 31.5 31.4 31.3 31.8 32.4 ...
## $ long  : num –78.3 –78.8 –79 –79.1 –79.2 –79.3 –79.4 –79 –78.7 –78.2
...
## $ status  : chr "tropical depression" "tropical depression" "tropical
depression" "tropical depression" ...
## $ category : Ord.factor w/ 7 levels "–1"<"0"<"1"<"2"<..: 1 1 1 1 2 2 2
2 2 2 ...
## $ wind  : int 25 25 25 30 35 35 40 50 50 60 ...
## $ pressure : int 1010 1009 1009 1009 1009 1007 1005 992 993 987 ...
## $ ts_diameter: num 0 0 0 0 57.5 ...
## $ hu_diameter: num 0 0 0 0 0 0 0 0 0 0 ...
head(dat)
## # A tibble: 6 x 13
## name year month day hour lat long status category wind pressure
## <chr> <dbl> <dbl> <int> <dbl> <dbl> <dbl> <chr> <ord> <int> <int>
## 1 Alex 2004  7 31 18 30.3 –78.3 tropi… –1   25  1010
## 2 Alex 2004  8  1  0 31 –78.8 tropi… –1   25  1009
## 3 Alex 2004  8  1  6 31.5 –79 tropi… –1   25  1009
## 4 Alex 2004  8  1 12 31.6 –79.1 tropi… –1   30  1009
```

```
## 5 Alex 2004  8  1 18 31.6 –79.2 tropi… 0   35  1009
## 6 Alex 2004  8  2  0 31.5 –79.3 tropi… 0   35  1007
## # … with 2 more variables: ts_diameter <dbl>, hu_diameter <dbl>
```

The subsetted data contains only 3453 rows for all storms except Katrina and Keith storms and no missing data in ts_diameter and hu_diameter columns.

### 10.3.4 Arrange

The arrange() function is used to reorder rows of a data frame according to one of the columns.

Example, arrange the storms data so that the first row is smallest wind speed and last row is maximum wind speed.

```
# arrange dataframe

dat<-arrange(storms, wind)

# first 6 rows
head(dat)
## # A tibble: 6 x 13
## name year month day hour lat long status category wind pressure
## <chr> <dbl> <dbl> <int> <dbl> <dbl> <dbl> <chr> <ord> <int> <int>
## 1 Bonn… 1986  6 28  6 36.5 –91.3 tropi… –1   10  1013
## 2 Bonn… 1986  6 28 12 37.2 –90 tropi… –1   10  1012
## 3 AL03… 1987  8 16 18 30.9 –83.2 tropi… –1   10  1014
## 4 AL03… 1987  8 17  0 31.4 –82.9 tropi… –1   10  1015
## 5 AL03… 1987  8 17  6 31.8 –82.3 tropi… –1   10  1015
## 6 Albe… 1994  7  7  0 32.7 –86.3 tropi… –1   10  1012
## # … with 2 more variables: ts_diameter <dbl>, hu_diameter <dbl>
# to see further deep in the data

head(dat,100)
```

```
## # A tibble: 100 x 13
## name year month day hour lat long status category wind pressure
## <chr> <dbl> <dbl> <int> <dbl> <dbl> <dbl> <chr> <ord> <int> <int>
## 1 Bonn… 1986 6 28 6 36.5 –91.3 tropi… –1 10 1013
## 2 Bonn… 1986 6 28 12 37.2 –90 tropi… –1 10 1012
## 3 AL03… 1987 8 16 18 30.9 –83.2 tropi… –1 10 1014
## 4 AL03… 1987 8 17 0 31.4 –82.9 tropi… –1 10 1015
## 5 AL03… 1987 8 17 6 31.8 –82.3 tropi… –1 10 1015
## 6 Albe… 1994 7 7 0 32.7 –86.3 tropi… –1 10 1012
## 7 Albe… 1994 7 7 6 32.7 –86.6 tropi… –1 10 1012
## 8 Albe… 1994 7 7 12 32.8 –86.8 tropi… –1 10 1012
## 9 Albe… 1994 7 7 18 33 –87 tropi… –1 10 1013
## 10 Clau… 1979 7 27 12 34 –95.9 tropi… –1 15 1007
## # … with 90 more rows, and 2 more variables: ts_diameter <dbl>,
## # hu_diameter <dbl>
# last 6 rows

tail(dat)
## # A tibble: 6 x 13
## name year month day hour lat long status category wind pressure
## <chr> <dbl> <dbl> <int> <dbl> <dbl> <dbl> <chr> <ord> <int> <int>
## 1 Mitch 1998 10 26 18 16.9 –83.1 hurri… 5 155 905
## 2 Mitch 1998 10 27 0 17.2 –83.8 hurri… 5 155 910
## 3 Rita 2005 9 22 3 24.7 –87.3 hurri… 5 155 895
## 4 Rita 2005 9 22 6 24.8 –87.6 hurri… 5 155 897
## 5 Gilb… 1988 9 14 0 19.7 –83.8 hurri… 5 160 888
## 6 Wilma 2005 10 19 12 17.3 –82.8 hurri… 5 160 882
## # … with 2 more variables: ts_diameter <dbl>, hu_diameter <dbl>
```

The arranged dat dataframe shows that the storms with minimum speed are Bonnie, AL031987, and Alberto with minimum wind speed of 10, while Gilbert and Wilma have the maximum wind speed of 160.

Data can be arranged in descending order by using the desc() operator.

```
# arrange dataframe in a descending order

dat<-arrange(storms, desc(wind))

# first 6 rows
head(dat)
## # A tibble: 6 x 13
## name year month day hour lat long status category wind pressure
## <chr> <dbl> <dbl> <int> <dbl> <dbl> <dbl> <chr> <ord> <int> <int>
## 1 Gilb… 1988 9 14 0 19.7 –83.8 hurri… 5 160 888
## 2 Wilma 2005 10 19 12 17.3 –82.8 hurri… 5 160 882
## 3 Gilb… 1988 9 14 6 19.9 –85.3 hurri… 5 155 889
## 4 Mitch 1998 10 26 18 16.9 –83.1 hurri… 5 155 905
## 5 Mitch 1998 10 27 0 17.2 –83.8 hurri… 5 155 910
## 6 Rita 2005 9 22 3 24.7 –87.3 hurri… 5 155 895
## # … with 2 more variables: ts_diameter <dbl>, hu_diameter <dbl>
# last 6 rows

tail(dat)
## # A tibble: 6 x 13
## name year month day hour lat long status category wind pressure
## <chr> <dbl> <dbl> <int> <dbl> <dbl> <dbl> <chr> <ord> <int> <int>
## 1 AL03… 1987 8 17 0 31.4 –82.9 tropi… –1 10 1015
## 2 AL03… 1987 8 17 6 31.8 –82.3 tropi… –1 10 1015
## 3 Albe… 1994 7 7 0 32.7 –86.3 tropi… –1 10 1012
## 4 Albe… 1994 7 7 6 32.7 –86.6 tropi… –1 10 1012
## 5 Albe… 1994 7 7 12 32.8 –86.8 tropi… –1 10 1012
## 6 Albe… 1994 7 7 18 33 –87 tropi… –1 10 1013
## # … with 2 more variables: ts_diameter <dbl>, hu_diameter <dbl>
```

Note that the arranged data has the same dimensions as the original data, 10010 rows and 13 columns

### 10.3.5 Rename

The rename() function is used to rename columns of a dataframe in an easy way. The syntax inside the rename() function is to have the new name on the left-hand side of the = sign and the old name on the right-hand side.

To rename the last as latitude and long as longitude of the storms dataframe

```
head(storms)
## # A tibble: 6 x 13
## name year month day hour lat long status category wind pressure
## <chr> <dbl> <dbl> <int> <dbl> <dbl> <dbl> <chr> <ord> <int> <int>
## 1 Amy 1975  6 27  0 27.5 –79 tropi… –1   25  1013
## 2 Amy 1975  6 27  6 28.5 –79 tropi… –1   25  1013
## 3 Amy 1975  6 27 12 29.5 –79 tropi… –1   25  1013
## 4 Amy 1975  6 27 18 30.5 –79 tropi… –1   25  1013
## 5 Amy 1975  6 28  0 31.5 –78.8 tropi… –1   25  1012
## 6 Amy 1975  6 28  6 32.4 –78.7 tropi… –1   25  1012
## # … with 2 more variables: ts_diameter <dbl>, hu_diameter <dbl>
storms<-rename(storms, latitude = lat, longitude = long)
```

```
head(storms)
## # A tibble: 6 x 13
## name year month day hour latitude longitude status category wind
## <chr> <dbl> <dbl> <int> <dbl> <dbl>  <dbl> <chr> <ord> <int>
## 1 Amy 1975  6 27  0  27.5  –79 tropi… –1   25
## 2 Amy 1975  6 27  6  28.5  –79 tropi… –1   25
## 3 Amy 1975  6 27 12  29.5  –79 tropi… –1   25
## 4 Amy 1975  6 27 18  30.5  –79 tropi… –1   25
## 5 Amy 1975  6 28  0  31.5  –78.8 tropi… –1   25
```

```
## 6 Amy 1975 6 28 6 32.4 –78.7 tropi… –1 25
## # … with 3 more variables: pressure <int>, ts_diameter <dbl>, hu_
diameter <dbl>
```

### 10.3.6 Mutate

The mutate() function can be used to add new columns to a data frame or change existing columns in the data frame.

There is also the related transmute() function, which does the same thing as mutate() but then drops all non-transformed variables.

We can add a date column to the storms dataframe using make_date() function of the lubridate package (see Chapter 8).

```
library(lubridate)
##
## Attaching package: 'lubridate'
## The following object is masked from 'package:base':
##
## date
head(storms)
## # A tibble: 6 x 13
## name year month day hour latitude longitude status category wind
## <chr> <dbl> <dbl> <int> <dbl> <dbl> <dbl> <chr> <ord> <int>
## 1 Amy 1975 6 27 0 27.5 –79 tropi… –1 25
## 2 Amy 1975 6 27 6 28.5 –79 tropi… –1 25
## 3 Amy 1975 6 27 12 29.5 –79 tropi… –1 25
## 4 Amy 1975 6 27 18 30.5 –79 tropi… –1 25
## 5 Amy 1975 6 28 0 31.5 –78.8 tropi… –1 25
## 6 Amy 1975 6 28 6 32.4 –78.7 tropi… –1 25
## # … with 3 more variables: pressure <int>, ts_diameter <dbl>, hu_
diameter <dbl>
storms<-mutate(storms, date = make_date(year = year, month = month,
day = day))
```

```
head(storms)
## # A tibble: 6 x 14
## name year month day hour latitude longitude status category wind
## <chr> <dbl> <dbl> <int> <dbl> <dbl>  <dbl> <chr> <ord> <int>
## 1 Amy 1975  6 27  0  27.5  –79 tropi… –1   25
## 2 Amy 1975  6 27  6  28.5  –79 tropi… –1   25
## 3 Amy 1975  6 27 12  29.5  –79 tropi… –1   25
## 4 Amy 1975  6 27 18  30.5  –79 tropi… –1   25
## 5 Amy 1975  6 28  0  31.5  –78.8 tropi… –1   25
## 6 Amy 1975  6 28  6  32.4  –78.7 tropi… –1   25
## # … with 4 more variables: pressure <int>, ts_diameter <dbl>,
## # hu_diameter <dbl>, date <date>
head(storms$date)
## [1] "1975-06-27" "1975-06-27" "1975-06-27" "1975-06-27" "1975-06-28"
## [6] "1975-06-28"
class(storms$date)
## [1] "Date"
```

In this mutate() function, we pass the columns, year, month, day of storms data to year, month, day arguments of the make_date() function

This column also can be created using the paste() function but its class will be character instead of date

```
head(storms)
## # A tibble: 6 x 14
## name year month day hour latitude longitude status category wind
## <chr> <dbl> <dbl> <int> <dbl> <dbl>  <dbl> <chr> <ord> <int>
## 1 Amy 1975  6 27  0  27.5  –79 tropi… –1   25
## 2 Amy 1975  6 27  6  28.5  –79 tropi… –1   25
## 3 Amy 1975  6 27 12  29.5  –79 tropi… –1   25
```

```
## 4 Amy 1975  6 27 18  30.5  –79 tropi… –1   25
## 5 Amy 1975  6 28  0  31.5  –78.8 tropi… –1   25
## 6 Amy 1975  6 28  6  32.4  –78.7 tropi… –1   25
## # … with 4 more variables: pressure <int>, ts_diameter <dbl>,
## # hu_diameter <dbl>, date <date>
storms<-mutate(storms, date = paste(year, month, day, sep = "-"))

head(storms)
## # A tibble: 6 x 14
## name year month day hour latitude longitude status category wind
## <chr> <dbl> <dbl> <int> <dbl> <dbl>  <dbl> <chr> <ord> <int>
## 1 Amy 1975  6 27  0  27.5  –79 tropi… –1   25
## 2 Amy 1975  6 27  6  28.5  –79 tropi… –1   25
## 3 Amy 1975  6 27 12  29.5  –79 tropi… –1   25
## 4 Amy 1975  6 27 18  30.5  –79 tropi… –1   25
## 5 Amy 1975  6 28  0  31.5  –78.8 tropi… –1   25
## 6 Amy 1975  6 28  6  32.4  –78.7 tropi… –1   25
## # … with 4 more variables: pressure <int>, ts_diameter <dbl>,
## # hu_diameter <dbl>, date <chr>
head(storms$date)
## [1] "1975-6-27" "1975-6-27" "1975-6-27" "1975-6-27" "1975-6-28"
"1975-6-28"
class(storms$date)
## [1] "character"
```

Using transmute() function

```
head(storms)
## # A tibble: 6 x 14
## name year month day hour latitude longitude status category wind
## <chr> <dbl> <dbl> <int> <dbl> <dbl>  <dbl> <chr> <ord> <int>
```

```
## 1 Amy 1975 6 27 0 27.5 –79 tropi… –1 25
## 2 Amy 1975 6 27 6 28.5 –79 tropi… –1 25
## 3 Amy 1975 6 27 12 29.5 –79 tropi… –1 25
## 4 Amy 1975 6 27 18 30.5 –79 tropi… –1 25
## 5 Amy 1975 6 28 0 31.5 –78.8 tropi… –1 25
## 6 Amy 1975 6 28 6 32.4 –78.7 tropi… –1 25
## # … with 4 more variables: pressure <int>, ts_diameter <dbl>,
## # hu_diameter <dbl>, date <chr>
storms<-transmute(storms, date = paste(year, month, day, sep = "-"))

head(storms)
## # A tibble: 6 x 1
## date
## <chr>
## 1 1975-6-27
## 2 1975-6-27
## 3 1975-6-27
## 4 1975-6-27
## 5 1975-6-28
## 6 1975-6-28
str(storms)
## Classes 'tbl_df', 'tbl' and 'data.frame': 10010 obs. of 1 variable:
## $ date: chr "1975-6-27" "1975-6-27" "1975-6-27" "1975-6-27" ...
data("storms")
```

The data now contains only one transformed column. To load the original data, use data(storms).

### 10.3.7 Pipeline Operator %>%

The pipeline operator %>% is very useful for stringing together multiple dplyr functions in a in a left-to-right fashion.

Example, consider the following steps of data analysis:

1. Filter for rows containing storm name "Katrina;"
2. Select name, year, wind, pressure columns;
3. Arrange the rows according to wind speed in descending order.

*Solution 1:* Nest many dplyr functions

```
dat<-arrange(select(filter(storms, name=="Katrina"), name,year,wind, pressure), desc(wind))

str(dat)
## Classes 'tbl_df', 'tbl' and 'data.frame': 68 obs. of 4 variables:
## $ name : chr "Katrina" "Katrina" "Katrina" "Katrina" ...
## $ year : num 2005 2005 2005 2005 2005 ...
## $ wind : int 150 145 140 125 125 110 110 105 100 100 ...
## $ pressure: int 902 909 905 930 913 920 923 928 942 948 ...
# first rows containing high wind speeds

head(dat)
## # A tibble: 6 x 4
## name  year wind pressure
## <chr> <dbl> <int> <int>
## 1 Katrina 2005 150  902
## 2 Katrina 2005 145  909
## 3 Katrina 2005 140  905
## 4 Katrina 2005 125  930
## 5 Katrina 2005 125  913
## 6 Katrina 2005 110  920
# last rows containing low wind speeds


tail(dat)
## # A tibble: 6 x 4
## name  year wind pressure
## <chr> <dbl> <int> <int>
```

```
## 1 Katrina 1999 25  1007
## 2 Katrina 1999 25  1008
## 3 Katrina 1999 25  1009
## 4 Katrina 1999 20  1010
## 5 Katrina 1999 20  1011
## 6 Katrina 1999 20  1011
```

The created dataframe consists of 68 rows and 4 columns. However, the multiple dplyr functions are nested in a manner that is difficult to read.

*Solution 2:* using pipeline operator %>% that connects functions in a left to right manner.

```
dat<-arrange(select(filter(storms, name=="Katrina"), name,year,wind, pressure), desc(wind))



dat2<- storms %>% filter(name=="Katrina") %>% select(name, year, wind, pressure) %>%

 arrange(desc(wind))

str(dat2)
## Classes 'tbl_df', 'tbl' and 'data.frame': 68 obs. of 4 variables:
## $ name : chr "Katrina" "Katrina" "Katrina" "Katrina" ...
## $ year : num 2005 2005 2005 2005 2005 ...
## $ wind : int 150 145 140 125 125 110 110 105 100 100 ...
## $ pressure: int 902 909 905 930 913 920 923 928 942 948 ...
head(dat2)
## # A tibble: 6 x 4
## name  year wind pressure
## <chr> <dbl> <int> <int>
## 1 Katrina 2005 150  902
```

```
## 2 Katrina 2005 145  909
## 3 Katrina 2005 140  905
## 4 Katrina 2005 125  930
## 5 Katrina 2005 125  913
## 6 Katrina 2005 110  920
tail(dat2)
## # A tibble: 6 x 4
## name  year wind pressure
## <chr> <dbl> <int> <int>
## 1 Katrina 1999 25  1007
## 2 Katrina 1999 25  1008
## 3 Katrina 1999 25  1009
## 4 Katrina 1999 20  1010
## 5 Katrina 1999 20  1011
## 6 Katrina 1999 20  1011
identical(dat,dat2)
## [1] TRUE
```

Using the pipeline operator %>% has generated the same dataframe but with simple-to-read one line of code.

### 10.3.8 Summarise/Summarize

summarise() takes a dataframe and creates a new data frame with the requested summaries. In conjunction with summarize, you can use any R functions to create this summary.

Example, to summarize the storms dataframe and create a dataframe containing the number of observations or rows (n() function), the maximum wind speed (max() function), the minimum pressure (min() function)

```
dat<-storms %>% summarise(observations = n(), max_wind = max(wind),
min_pressure = min(pressure))
dat
## # A tibble: 1 x 3
## observations max_wind min_pressure
```

```
## <int> <int> <int>
## 1 10010 160 882
str(dat)
## Classes 'tbl_df', 'tbl' and 'data.frame': 1 obs. of 3 variables:
## $ observations: int 10010
## $ max_wind : int 160
## $ min_pressure: int 882
```

The new dataframe consists of 3 columns (observations, max_wind, min_pressure) and 1 row containing the required values. The number of observations is 10010, the maximum wind speed = 160, and the minimum pressure = 882.

Using summarise() with group_by() function will create the dataframe with summary statistics for each stratum.

Example, group the storms dataframe by month and then summarize the grouped storms to get the number of observations or rows (n() function), the maximum wind speed (max() function), the minimum pressure (min() function) for each month.

```
dat<-storms %>% group_by(month) %>% summarise(observations =
n(),

        max_wind = max(wind),

        min_pressure = min(pressure))

dat
## # A tibble: 10 x 4
## month observations max_wind min_pressure
## <dbl> <int> <int> <int>
## 1 1 23 55 994
## 2 4 13 50 994
## 3 5 53 60 992
```

```
## 4 6  310 80  958
## 5 7  774 140  929
## 6 8  2400 150  902
## 7 9  4120 160  888
## 8 10  1637 160  882
## 9 11  555 135  933
## 10 12  125 75  979
str(dat)
## Classes 'tbl_df', 'tbl' and 'data.frame': 10 obs. of 4 variables:
## $ month  : num 1 4 5 6 7 8 9 10 11 12
## $ observations: int 23 13 53 310 774 2400 4120 1637 555 125
## $ max_wind : int 55 50 60 80 140 150 160 160 135 75
## $ min_pressure: int 994 994 992 958 929 902 888 882 933 979
```

The new dataframe consists of 10 rows (month 2,3 are not present in storms dataframe) and 4 columns:

1. Month: the grouping column.
2. Observations column contains the number of rows for each month.
3. max_wind column contains the maximum wind speed for each month.
4. min_pressure column contains the minimum pressure for each month.

To go a step further in your data analysis, you can arrange the dataframe according to maximum wind in a descending order to see quickly which month had the maximum wind storms.

```
dat<-storms %>% group_by(month) %>% summarise(observations =
n(),

          max_wind = max(wind),

          min_pressure = min(pressure)) %>%
```

```
arrange(desc(max_wind))


dat
## # A tibble: 10 x 4
## month observations max_wind min_pressure
## <dbl>  <int> <int>  <int>
## 1  9   4120  160   888
## 2 10   1637  160   882
## 3  8   2400  150   902
## 4  7   774  140   929
## 5 11   555  135   933
## 6  6   310  80   958
## 7 12   125  75   979
## 8  5   53  60   992
## 9  1   23  55   994
## 10  4   13  50   994
```

The maximum wind storms has occurred in months 9 and 10 (September and October).

You can also group by many columns to get many summary statistics.

*Example:* Group the storms dataframe by name and year and summarize to get maximum wind speed then arrange by that maximum wind speed.

```
dat<-storms %>% group_by(name, year) %>% summarise(max_wind = max(wind)) %>%

arrange(desc(max_wind))

head(dat)
## # A tibble: 6 x 3
```

```
## # Groups: name [6]
## name  year max_wind
## <chr> <dbl> <int>
## 1 Gilbert 1988  160
## 2 Wilma 2005  160
## 3 Mitch 1998  155
## 4 Rita  2005  155
## 5 Andrew 1992  150
## 6 Anita 1977  150
tail(dat)
## # A tibble: 6 x 3
## # Groups: name [5]
## name  year max_wind
## <chr> <dbl> <int>
## 1 Two  2010  30
## 2 Two  2014  30
## 3 AL012000 2000  25
## 4 AL022001 2001  25
## 5 AL101991 1991  25
## 6 Sixteen 2008  25
dim(dat)
## [1] 426 3
```

The new dataframe consists of 426 rows and 3 columns (name, year, max_wind)

From the head() and tail () functions:

1. The maximum wind speed storms (for this dataframe) were Gilbert occurred in 1988 and Wilma occurred in 2005.
2. The minimum wind speed storms (for this dataframe) were Sixteen occurred in 2008, AL101991 occurred in 1991, AL022001 occurred in 2001, and AL012000 occurred in 2000.

## 10.4 TIDYR PACKAGE

The tidyr package includes functions to transfer a data frame between long and wide formats. Wide format data have different variables describing an observation placed in separate columns. Long format data have different attributes encoded as levels of a single variable, followed by another column that contains the values of the observation at those different levels. In R for data analysis, it is preferred to have the data in long format.

### 10.4.1 gather()

The gather() function is used to gather values spread across several columns into a single column, with the column names gathered into a "key" column and values gathered into a "value" column. When gathering, exclude any columns that you don't want to gather by including the column names with a minus sign in the gather function.

Example:

VADeaths is a data set for the Death rates per 1000 in Virginia in 1940. It is a matrix of 5 rows and 4 columns

**data**("VADeaths")

VADeaths

## Rural Male Rural Female Urban Male Urban Female

## 50–54 11.7 8.7 15.4 8.4

## 55–59 18.1 11.7 24.3 13.6

## 60–64 26.9 20.3 37.0 19.3

## 65–69 41.0 30.9 54.6 35.1

## 70–74 66.0 54.3 71.1 50.0

The data is not tidy because there are variables in both the rows and columns. The variables are age category, gender, and urbanness, and the death rate which is presented inside the table.

The first step is converting matrix to dataframe and convert the row names to another column in that dataframe using mutate() function or rownames_to_column() function

VADeaths **%>%** **data.frame()** **%>%** **mutate**(age_category = **rownames**(VADeaths))

```
## Rural.Male Rural.Female Urban.Male Urban.Female age_category
## 1 11.7 8.7 15.4 8.4 50–54
## 2 18.1 11.7 24.3 13.6 55–59
## 3 26.9 20.3 37.0 19.3 60–64
## 4 41.0 30.9 54.6 35.1 65–69
## 5 66.0 54.3 71.1 50.0 70–74
```

VADeaths **%>%** **data.frame()** **%>%** **rownames_to_column**("age_category")

```
## age_category Rural.Male Rural.Female Urban.Male Urban.Female
## 1 50–54 11.7 8.7 15.4 8.4
## 2 55–59 18.1 11.7 24.3 13.6
## 3 60–64 26.9 20.3 37.0 19.3
## 4 65–69 41.0 30.9 54.6 35.1
## 5 70–74 66.0 54.3 71.1 50.0
```

In both cases, the rownames are converted to a column named age_category and the data is now a dataframe with 5 rows and 5 columns

The second step is gathering all columns to a column called factor (that is a factor with levels, Rural.Male Rural.Female Urban.Male Urban.Female) and the values in another column called death_rate. We exclude age_category column from this gathering.

VADeaths **%>%** **data.frame()** **%>%** **rownames_to_column**("age_category") **%>%**

**gather**(key = "factor", value = "death_rate", -age_category)

```
## age_category factor death_rate
## 1 50–54 Rural.Male 11.7
## 2 55–59 Rural.Male 18.1
## 3 60–64 Rural.Male 26.9
## 4 65–69 Rural.Male 41.0
```

```
## 5  70–74 Rural.Male  66.0
## 6  50–54 Rural.Female  8.7
## 7  55–59 Rural.Female  11.7
## 8  60–64 Rural.Female  20.3
## 9  65–69 Rural.Female  30.9
## 10  70–74 Rural.Female  54.3
## 11  50–54 Urban.Male  15.4
## 12  55–59 Urban.Male  24.3
## 13  60–64 Urban.Male  37.0
## 14  65–69 Urban.Male  54.6
## 15  70–74 Urban.Male  71.1
## 16  50–54 Urban.Female  8.4
## 17  55–59 Urban.Female  13.6
## 18  60–64 Urban.Female  19.3
## 19  65–69 Urban.Female  35.1
## 20  70–74 Urban.Female  50.0
```

The data is now a dataframes with 20 rows and 3 columns (age_category, factor, death_rate).

According to R help page for gather() function, it is recommended to use pivot_longer() function because it is easier to use, more featureful, and still under active development. df %>% gather("key", "value", x, y, z) is equivalent to df %>% pivot_longer(c(x, y, z), names_to = "key", values_to = "value"). x,y,z are columns to gather and if not named, the gather() or pivot_longer() function will gather all columns.

```
VADeaths %>% data.frame() %>% rownames_to_column("age_category") %>%

  pivot_longer(names_to = "factor", values_to = "death_rate", -age_category)
## # A tibble: 20 x 3
## age_category factor  death_rate
```

```
## <chr> <chr>   <dbl>
## 1 50–54  Rural.Male   11.7
## 2 50–54  Rural.Female  8.7
## 3 50–54  Urban.Male   15.4
## 4 50–54  Urban.Female  8.4
## 5 55–59  Rural.Male   18.1
## 6 55–59  Rural.Female  11.7
## 7 55–59  Urban.Male   24.3
## 8 55–59  Urban.Female  13.6
## 9 60–64  Rural.Male   26.9
## 10 60–64  Rural.Female  20.3
## 11 60–64  Urban.Male   37
## 12 60–64  Urban.Female  19.3
## 13 65–69  Rural.Male   41
## 14 65–69  Rural.Female  30.9
## 15 65–69  Urban.Male   54.6
## 16 65–69  Urban.Female  35.1
## 17 70–74  Rural.Male   66
## 18 70–74  Rural.Female  54.3
## 19 70–74  Urban.Male   71.1
## 20 70–74  Urban.Female  50
```

The same dataframe is produced with 20 rows and 3 columns (age_category, factor, death_rate).

### 10.4.2 Separate()

separate() function is used to turn a single character column into multiple columns.

Although the last produced dataframe, of 20 rows and 3 columns, is in tidy format, we notice that the factor column contain two variables, gender, urbanness so we can separate it into two columns.

VADeaths **%>% data.frame() %>% rownames_to_column**("age_category") **%>%**

```
pivot_longer(names_to = "factor", values_to = "death_rate", -age_category) %>%

separate(col = "factor", into = c("urban","gender"))
## # A tibble: 20 x 4
## age_category urban gender death_rate
## <chr>  <chr> <chr>  <dbl>
## 1 50–54  Rural Male   11.7
## 2 50–54  Rural Female  8.7
## 3 50–54  Urban Male   15.4
## 4 50–54  Urban Female  8.4
## 5 55–59  Rural Male   18.1
## 6 55–59  Rural Female  11.7
## 7 55–59  Urban Male   24.3
## 8 55–59  Urban Female  13.6
## 9 60–64  Rural Male   26.9
## 10 60–64  Rural Female  20.3
## 11 60–64  Urban Male   37
## 12 60–64  Urban Female  19.3
## 13 65–69  Rural Male   41
## 14 65–69  Rural Female  30.9
## 15 65–69  Urban Male   54.6
## 16 65–69  Urban Female  35.1
## 17 70–74  Rural Male   66
## 18 70–74  Rural Female  54.3
## 19 70–74  Urban Male   71.1
## 20 70–74  Urban Female  50
```

Here we separated the factor column into two columns, "urdan" and "gender" so the dataframe now contains 20 rows and 4 columns.

### 10.4.3 Spread()

The spread() function is less commonly needed to tidy data but can be useful for creating summary tables.

The spread() function is the inverse of gather() function. According to R help page of spread(), development on spread() is complete, and for new code we recommend switching to pivot_wider(), which is easier to use, more featureful, and still under active development. df %>% spread(key, value) is equivalent to df %>% pivot_wider(names_from = key, values_from = value) where df is a dataframe.

The spread() function is used when you group the data by 2 columns then use spread() to convert one of the columns to separate columns.

*Example 1*: for storms data, create a summary table of the maximum wind speed for each year (last 5 years) and for each month. The data contains information from 1975–2015 so we filter for year > 2010.

We also use kable() function from knitr package to make a simple table with align argument = “c” to center cell values.

```
library(knitr)

# data before spread

storms %>% filter(year >2010) %>% group_by(year, month) %>%
summarise(max_wind= max(wind, na.rm = TRUE))
## # A tibble: 27 x 3
## # Groups: year [5]
##  year month max_wind
## <dbl> <dbl> <int>
## 1 2011  7  45
## 2 2011  8  60
## 3 2011  9  120
## 4 2011 10  120
## 5 2011 11  55
## 6 2012  5  60
```

```
## 7 2012  6  75
## 8 2012  8  95
## 9 2012  9  100
## 10 2012 10  100
## # … with 17 more rows
```

*# data with spread*

storms **%>% filter**(year >2010) **%>% group_by**(year, month) **%>% summarise**(max_wind= **max**(wind, na.rm = TRUE)) **%>% spread**(month, max_wind) **%>% kable**(align = "c", format = "markdown")

| Year | 5 | 6 | 7 | 8 | 9 | 10 | 11 |
|---|---|---|---|---|---|---|---|
| **2011** | NA | NA | 45 | 60 | 120 | 120 | 55 |
| **2012** | 60 | 75 | NA | 95 | 100 | 100 | NA |
| **2013** | NA | 55 | 55 | 50 | 80 | 55 | 55 |
| **2014** | NA | NA | 85 | 75 | 105 | 125 | NA |
| **2015** | 50 | 50 | 45 | 110 | 80 | 135 | 75 |

# data with pivot_wider

storms **%>% filter**(year >2010) **%>% group_by**(year, month) **%>% summarise**(max_wind= **max**(wind, na.rm = TRUE)) **%>% pivot_wider**(names_from = "month", values_from = max_wind) **%>% kable**(align = "c", format = "markdown")

| Year | 7 | 8 | 9 | 10 | 11 | 5 | 6 |
|---|---|---|---|---|---|---|---|
| **2011** | 45 | 60 | 120 | 120 | 55 | NA | NA |
| **2012** | NA | 95 | 100 | 100 | NA | 60 | 75 |
| **2013** | 55 | 50 | 80 | 55 | 55 | NA | 55 |
| **2014** | 85 | 75 | 105 | 125 | NA | NA | NA |
| **2015** | 45 | 110 | 80 | 135 | 75 | 50 | 50 |

Note here that we spread month column into different columns. NA values because there are no available data for the corresponding year and month.

*Example 2:* for regicor data from the package compareGroups, create a summay table of the maximum total cholesterol for each year and for each gender. The data contains information from 3 years 1995, 2000, 2005.

**Library**(compareGroups)

**data**("regicor")

*# data before spread*

regicor **%>% group_by**(year, sex) **%>% summarise**(max_cholesterol = **max**(chol, na.rm = TRUE))

```
## # A tibble: 6 x 3
## # Groups: year [3]
## year sex max_cholesterol
## <fct> <fct>   <dbl>
## 1 1995 Male    346
## 2 1995 Female    361
## 3 2000 Male    404
## 4 2000 Female    386
## 5 2005 Male    369
## 6 2005 Female    488
```

*# data after spread*

regicor **%>% group_by**(year, sex) **%>% summarise**(max_cholesterol = **max**(chol, na.rm = TRUE)) **%>% spread**(sex, max_cholesterol) **%>% kable**(align = "c", format = "markdown")

| Year | Male | Female |
|---|---|---|
| **1995** | 346 | 361 |
| **2000** | 404 | 386 |
| **2005** | 369 | 488 |

## 10.4.4 Unite()

unite() function is the complement of separate() function and is used to paste together multiple columns into one.

Example, we can create a column for gender_smoking of the regicor dataframe and then found maximum total cholesterol for each level

```
library(compareGroups)

data("regicor")

summary(regicor)
## id   year   age   sex
## Min. :2.000e+00 1995: 431 Min. :35.00 Male :1101
## 1st Qu.:2.128e+03 2000: 786 1st Qu.:46.00 Female:1193
## Median :1.000e+09 2005:1077 Median :55.00
## Mean :1.216e+09    Mean :54.74
## 3rd Qu.:3.000e+09    3rd Qu.:64.00
## Max. :3.000e+09    Max. :74.00
##
##      smoker   sbp   dbp   histhtn
## Never smoker    :1201 Min. : 80.0 Min. : 40.00 Yes : 723
## Current or former < 1y: 593 1st Qu.:116.0 1st Qu.: 72.00 No :1563
## Former >= 1y    : 439 Median :129.0 Median : 80.00 NA's: 8
## NA's     : 61 Mean :131.2 Mean : 79.66
##       3rd Qu.:144.0 3rd Qu.: 86.00
##       Max. :229.0 Max. :123.00
##       NA's :14  NA's :14
## txhtn   chol   hdl   triglyc   ldl
## No :1823 Min. : 95.0 Min. : 19.58 Min. : 25.0 Min. : 36.3
## Yes : 428 1st Qu.:189.0 1st Qu.: 42.00 1st Qu.: 72.0 1st Qu.:115.8
## NA's: 43 Median :215.0 Median : 51.00 Median : 97.0 Median :140.6
##    Mean :218.8 Mean : 52.69 Mean :115.6 Mean :143.2
##    3rd Qu.:245.0 3rd Qu.: 61.43 3rd Qu.:136.0 3rd Qu.:168.1
##    Max. :488.0 Max. :112.00 Max. :960.0 Max. :329.6
##    NA's :101  NA's :69  NA's :63  NA's :168
## histchol  txchol   height   weight   bmi
```

```
## Yes : 709 No :2011 Min. :137.0 Min. : 41.20 Min. :17.15
## No :1564 Yes : 228 1st Qu.:156.0 1st Qu.: 63.50 1st Qu.:24.38
## NA's: 21 NA's: 55 Median :162.5 Median : 73.00 Median :27.18
##      Mean :162.9 Mean : 73.44 Mean :27.64
##      3rd Qu.:169.2 3rd Qu.: 82.00 3rd Qu.:30.41
##      Max. :199.0 Max. :127.20 Max. :48.24
##      NA's :35  NA's :35  NA's :35
##  phyact   pcs    mcs   cv
## Min. : 0.0 Min. :13.95 Min. : 3.424 No :2071
## 1st Qu.: 159.5 1st Qu.:45.01 1st Qu.:42.181 Yes : 92
## Median : 303.7 Median :52.27 Median :51.260 NA's: 131
## Mean : 398.8 Mean :49.62 Mean :47.983
## 3rd Qu.: 521.9 3rd Qu.:55.78 3rd Qu.:56.047
## Max. :5083.2 Max. :67.06 Max. :69.904
## NA's :88  NA's :240  NA's :240
##  tocv   death   todeath
## Min. : 0.115 No :1975 Min. : 0.22
## 1st Qu.: 782.710 Yes : 173 1st Qu.: 787.63
## Median :1718.026 NA's: 146 Median :1668.40
## Mean :1754.668    Mean :1721.31
## 3rd Qu.:2690.549    3rd Qu.:2662.54
## Max. :3650.674    Max. :3651.26
## NA's :131     NA's :146
# there are NA values for smoker so we exclude it

dat<-regicor %>% filter(!is.na(smoker)) %>% unite("gender_smoker",
c(sex,smoker))

table(dat$gender_smoker)
##
## Female_Current or former < 1y   Female_Former >= 1y
```

```
## 183 79
## Female_Never smoker Male_Current or former < 1y
## 900 410
## Male_Former >= 1y Male_Never smoker
## 360 301
# to obtain summary statistic

dat %>% group_by(gender_smoker) %>% summarise(max_chol = max(chol, na.rm = TRUE)) %>% arrange(desc(max_chol))
## # A tibble: 6 x 2
## gender_smoker max_chol
## <chr> <dbl>
## 1 Female_Never smoker 488
## 2 Female_Current or former < 1y 412
## 3 Male_Never smoker 404
## 4 Male_Former >= 1y 372
## 5 Male_Current or former < 1y 364
## 6 Female_Former >= 1y 310
```

The maximum value of total cholesterol was for a female who have never smoked before.

CHAPTER 11

# DATA VISUALIZATION USING GGPLOT2

## CONTENTS

# 11.1 INTRODUCTION

The ggplot2 package (another member of the tidyverse) allows you to quickly visualize and explore data. In ggplot2, it is recommended that everything you want to plot is included in a dataframe as a column, and the first argument to ggplot must be a dataframe.

The basic steps to create a plot with ggplot2 are:

1. Create an object of the ggplot class using ggplot() function.
2. Add on geoms and other elements to create and customize the plot using +.

## 11.1.1 Create a ggplot Object

The ggplot() function will create a blank plot without anything in it. Instead, it typically specifies the data frame you want to use (first argument) and which columns of that data frame are mapped to which aesthetics.

Example from regicor data of compareGroups package, to plot a scatterplot between weight on x-axis and height on y-axis.

```
library(tidyverse)
## -- Attaching packages ------------------------------------------------------
-------- tidyverse 1.3.0 --
## <U+2713> ggplot2 3.2.1  <U+2713> purrr 0.3.3
## <U+2713> tibble 2.1.3  <U+2713> dplyr 0.8.3
## <U+2713> tidyr 1.0.0  <U+2713> stringr 1.4.0
## <U+2713> readr 1.3.1  <U+2713> forcats 0.4.0
## -- Conflicts ---------------------------------------------------------------
tidyverse_conflicts() --
## x dplyr::filter() masks stats::filter()
## x dplyr::lag() masks stats::lag()
library(compareGroups)
## Warning: package 'compareGroups' was built under R version 3.6.3
## Loading required package: SNPassoc
## Warning: package 'SNPassoc' was built under R version 3.6.3
## Loading required package: haplo.stats
```

```
## Warning: package 'haplo.stats' was built under R version 3.6.3
## Loading required package: survival
## Warning: package 'survival' was built under R version 3.6.3
## Loading required package: mvtnorm
## Loading required package: parallel
## Registered S3 method overwritten by 'SNPassoc':
## method   from
## summary.haplo.glm haplo.stats
data(regicor)

ggplot(data = regicor, aes(x=weight, y = height))
```

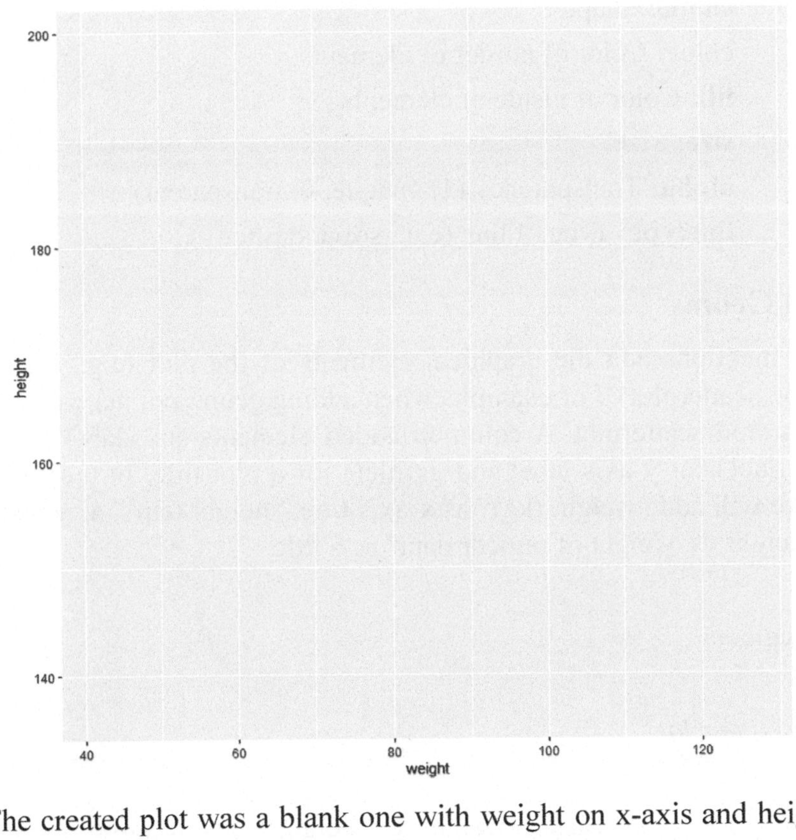


The created plot was a blank one with weight on x-axis and height on y-axis.

### 11.1.2 Aesthetics

Aesthetics are properties of the plot that can show certain elements of the data. For example, in the above plot, x-axis shows weight and y-axis shows height of the regicor dataframe. Which aesthetics are required for a plot depend on which geoms (plot types) you're adding to the plot. You can find out the aesthetics required for a geom in the "Aesthetics" section of the geom's help file (e.g., ?geom_point). Required aesthetics are in bold in this section of the help file and optional ones are not. The geom_point requires x and y aesthetics.

Common plot aesthetics you might want to specify include:

1. **x:** Position on x-axis
2. **y:** Position on y-axis
3. **shape:** Shape
4. **color:** Color of border of elements
5. **fill:** Color of inside of elements
6. **size:** Size
7. **alpha:** Transparency (1: opaque; 0: transparent)
8. **linetype:** Type of line (e.g., solid, dashed)

### 11.1.3 Geoms

Geom functions add the graphical elements of the plot (e.g., histogram, boxplot, scatterplot). For example, when adding geom_point(), you can plot the required scatterplot. A common added elements are xlab() for x-axis label, ylab() for y-axis label and ggtitle() for a plot title. In the following plot, we will add "weight (kg)" as x-axis title, "height (cm)" as y-axis title, and "height vs. weight of regicor data" as a title.

```
data(regicor)

# plot scatterplot

ggplot(data = regicor, aes(x=weight, y = height))+geom_point()
## Warning: Removed 35 rows containing missing values (geom_point).
```

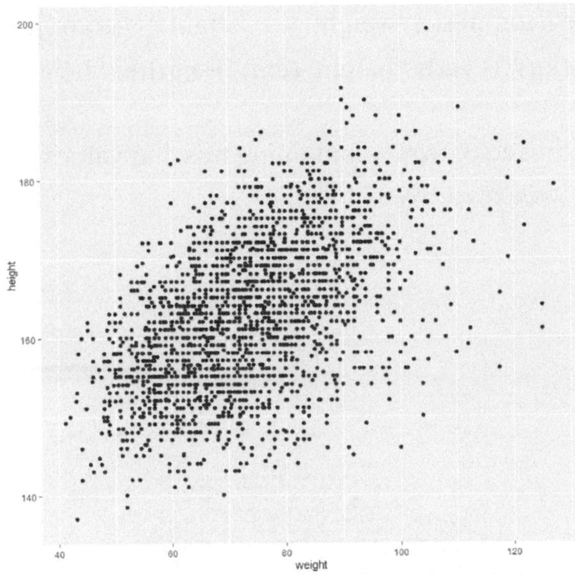


*# you can also put all arguments inside geom_point() function*

**ggplot**()+**geom_point**(data = regicor, **aes**(x=weight, y = height))
## Warning: Removed 35 rows containing missing values (geom_point).

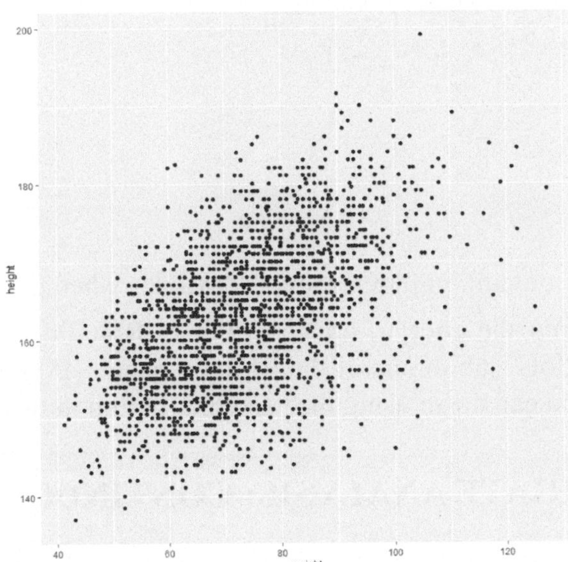


# Adding titles

**ggplot**(data = regicor, **aes**(x=weight, y = height))+**geom_point**()+
**xlab**("weight (kg)")+**ylab**("height (cm)")+**ggtitle**("height vs. weight of regicor data")

## Warning: Removed 35 rows containing missing values (geom_point).

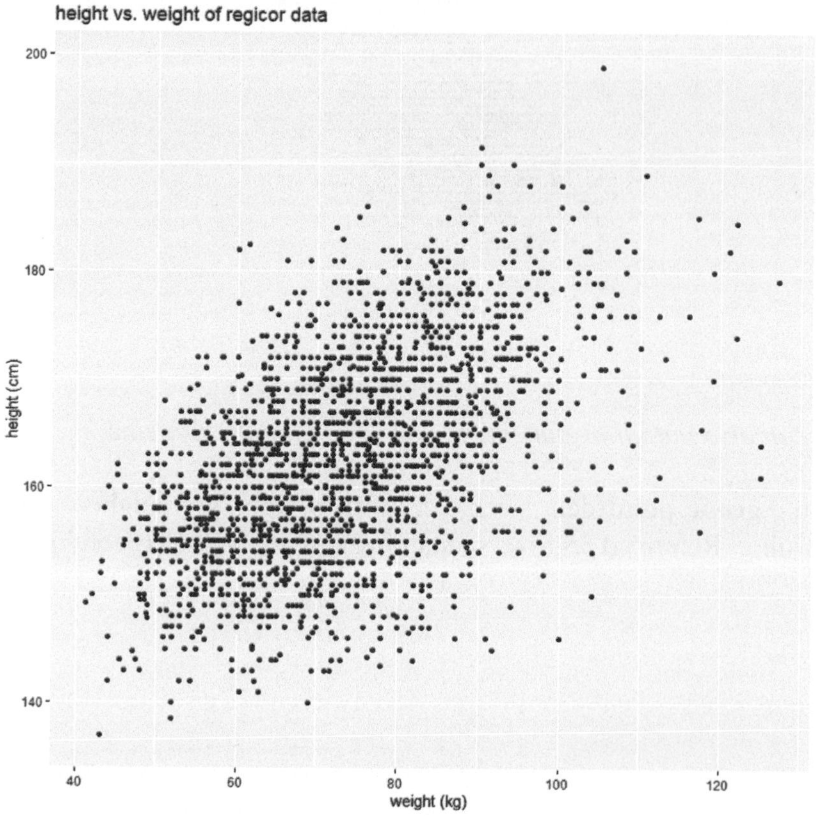


Note that by default, ggplot remove NA values when plotting data.

When you run the code to create a plot in RStudio, the plot will be shown in the "Plots" tab in one of the RStudio panels. If you would like to save the plot, you can do so using the "Export" button in this tab.

## 11.2 UNIVARIATE ANALYSIS: CONTINUOUS DATA

### 11.2.1 geom_histogram()

geom_histogram() is used to plot histograms. Histograms are used to visualize the distribution of a single continuous variable by dividing the x axis into bins, counting the number of observations in each bin, and displaying the counts with bars.

geom_histogram() requires only one aesthetic (x) or the continuous variable you want to visualize. One useful argument is bins which default to 30 bins.

Example, plot a histogram of regicor weight

data("regicor")

**ggplot**(data = regicor, **aes**(x = weight))+**geom_histogram**()

## `stat_bin()` using `bins = 30`. Pick better value with `binwidth`.

## Warning: Removed 35 rows containing non-finite values (stat_bin).

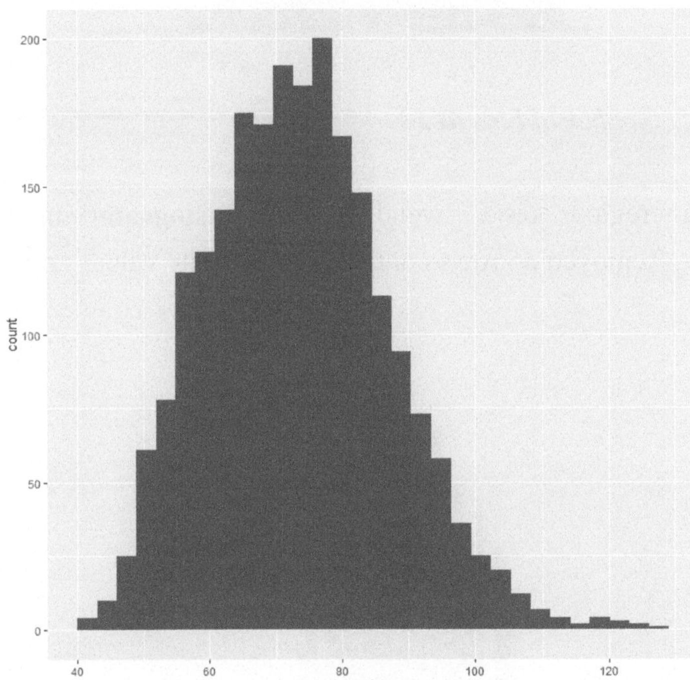


## change the number of bins to 20

**ggplot**(data = regicor, **aes**(x = weight))+**geom_histogram**(bins = 20)

## Warning: Removed 35 rows containing non-finite values (stat_bin).

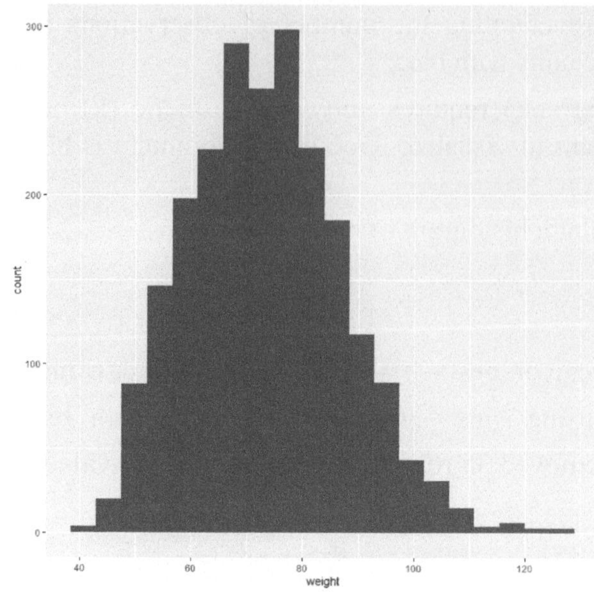


*# change the number of bins to 50*

**ggplot**(data = regicor, **aes**(x = weight))+**geom_histogram**(bins = 50)
## Warning: Removed 35 rows containing non-finite values (stat_bin).

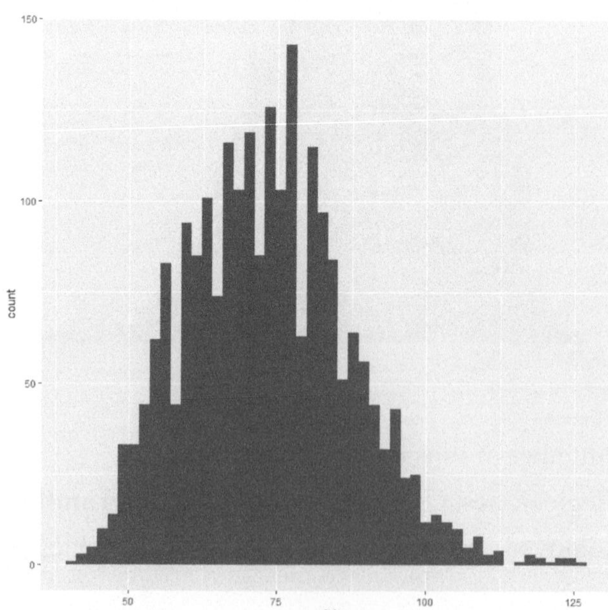

### 11.2.2 geom_boxplot()

geom_boxplot() is used to plot boxplots. Boxplots are used to compactly displays the distribution of a continuous variable. It visualizes five summary statistics (the median, first, and third quartile as box boundary), and all "outlying" points individually from which you can determine the minimum and maximum.

geom_boxplot() requires two aesthetics (y for the continuous variable you want to visualize on y-axis and group = 1 to indicate that all points are within a single group).

Example, plot a boxplot of regicor weight.

```
data("regicor")

ggplot(data = regicor, aes(y = weight, group=1))+geom_boxplot()
## Warning: Removed 35 rows containing non-finite values (stat_boxplot).
```

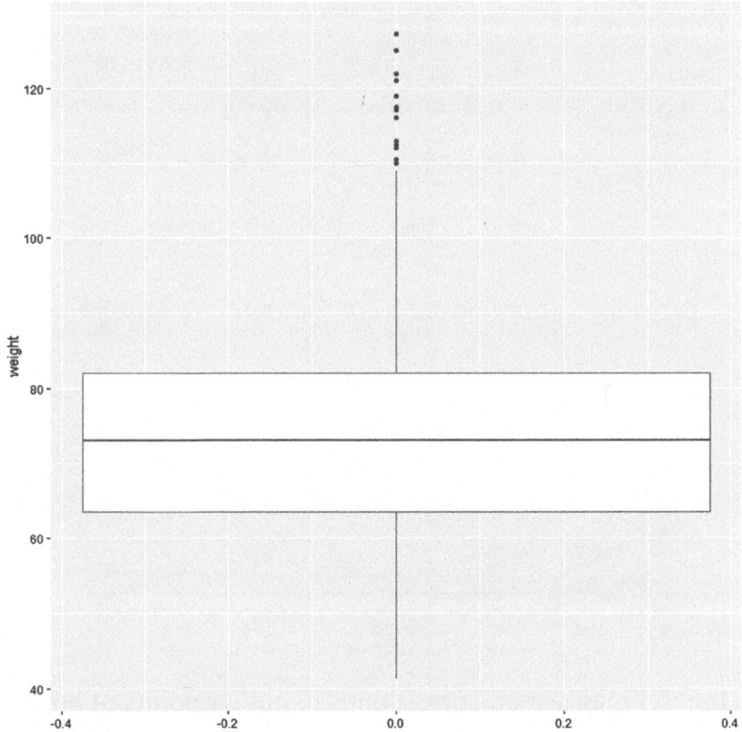

Note the outlier large points.

### 11.2.3 geom_density()

geom_density() is used to plot kernel density estimate. kernel density estimate is a smoothed version of the histogram. This is a useful alternative to the histogram for continuous data that comes from an underlying smooth distribution. The default kernel used is the gaussian one.

geom_density() requires only one aesthetic (x) or the continuous variable you want to visualize.

Example, plot a density plot of regicor weight.

```
data("regicor")

ggplot(data = regicor, aes(x = weight))+geom_density()
## Warning: Removed 35 rows containing non-finite values (stat_density).
```

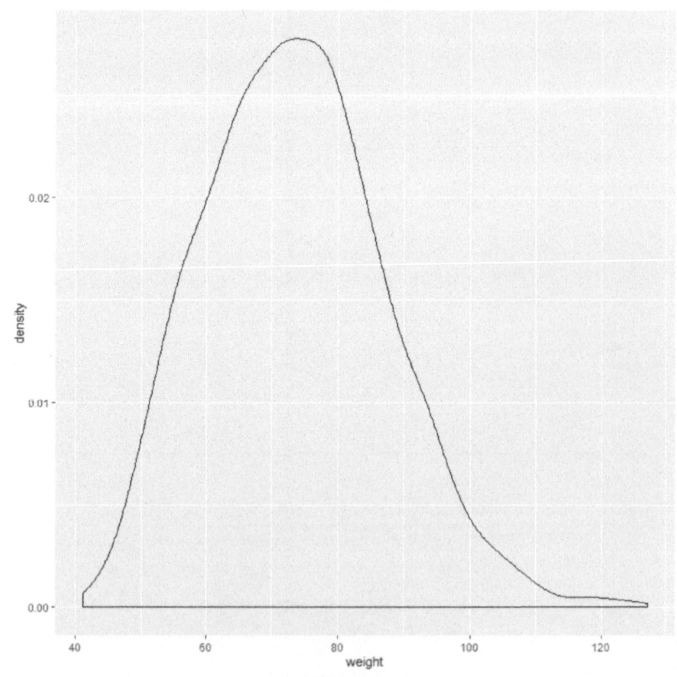


Note the right tail which corresponds to outlier points of boxplot.

### 11.2.4 geom_violin()

geom_violin() is used to plot violin plot. A violin plot is a compact display of a continuous distribution that is a mirrored density plot displayed in the same way as a boxplot.

geom_violin() requires two aesthetics (x for locating the symmetry line of the violin plot and y for the continuous variable you want to visualize on y-axis).

Example, plot a violin plot of regicor weight.

**data**("regicor")

**ggplot**(data = regicor, **aes**(x=1, y = weight))+**geom_violin**()

## Warning: Removed 35 rows containing non-finite values (stat_ydensity).

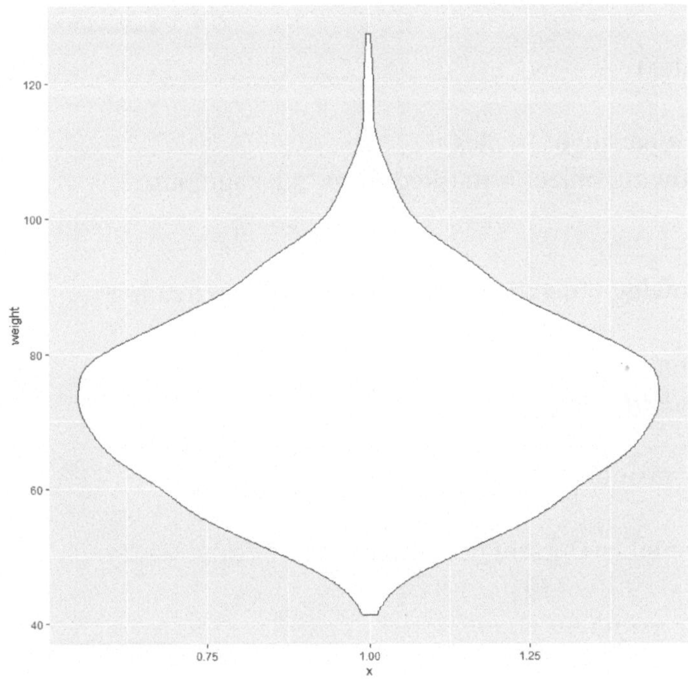


Note the tail at large points.

## 11.3 UNIVARIATE ANALYSIS: CATEGORICAL DATA

### 11.3.1 geom_bar() for Counts

geom_bar() is used to plot a bar graph where each bar has a height equal to the number of rows or observations at each level of the categorical variable. geom_bar() requires only one aesthetic (x) or the categorical variable you want to visualize. You can add another aesthetic, fill = categorical variable so every level will take a different color. To decrease the width of bars, use the argument width.

To get percentages, as that from prop.table(table()) function, it requires two aesthetics, x to be a constant like 1 (meaning that all data points are from 1 group) and fill = categorical variable. Also the position argument of geom_bar is set to "fill". We have also added another layer using scale_y_continuous() to convert the count to percentage format using percent_format() function of scales package.

Example, bar graph of smoker levels

```
library(scales)
##
## Attaching package: 'scales'
## The following object is masked from 'package:purrr':
##
## discard
## The following object is masked from 'package:readr':
##
## col_factor
# to get count of smoker levels

table(regicor$smoker)
##
## Never smoker Current or former < 1y Former >= 1y
## 1201 593 439
# to plot bar graph for counts

ggplot(data = regicor, aes(x= smoker))+geom_bar()
```

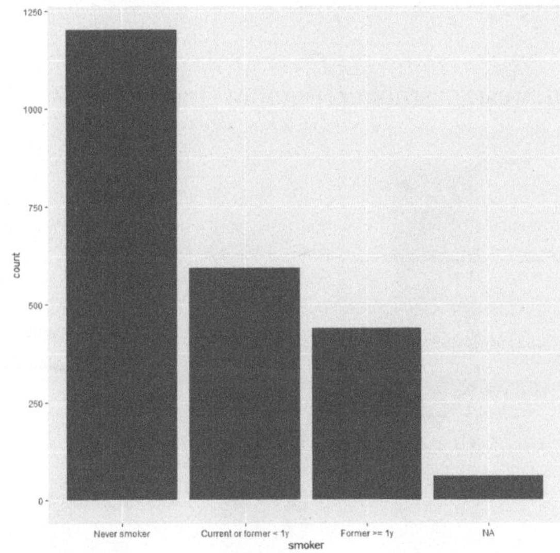


*# to remove NA values*

dat<-**filter**(regicor, **!is.na**(smoker))

**ggplot**(data = dat, **aes**(x= smoker))+**geom_bar**()

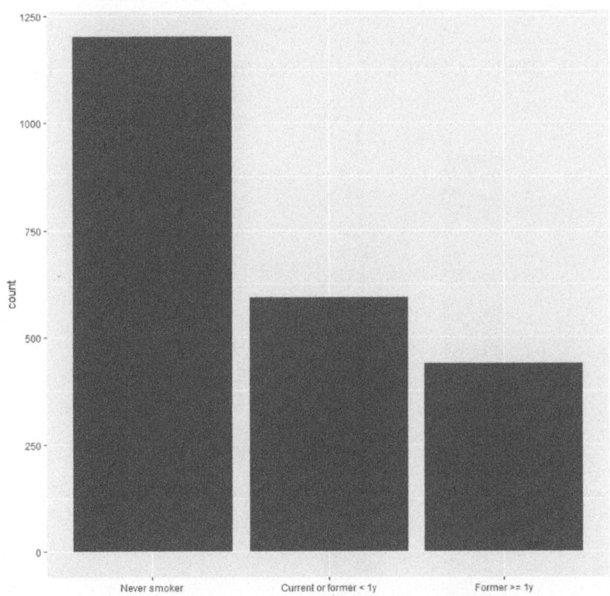

*# to reduce bar widths*

**ggplot**(data = dat, **aes**(x= smoker))+**geom_bar**(width = 0.5)

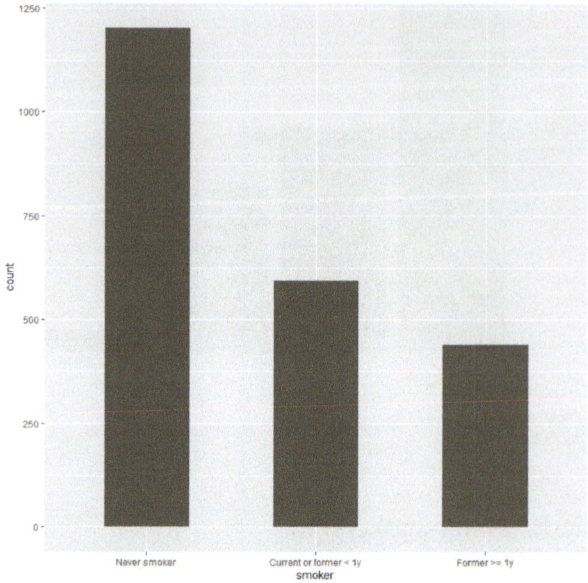


*# to add color to each level*

**ggplot**(data = dat, **aes**(x= smoker, fill = smoker))+**geom_bar**(width = 0.5)

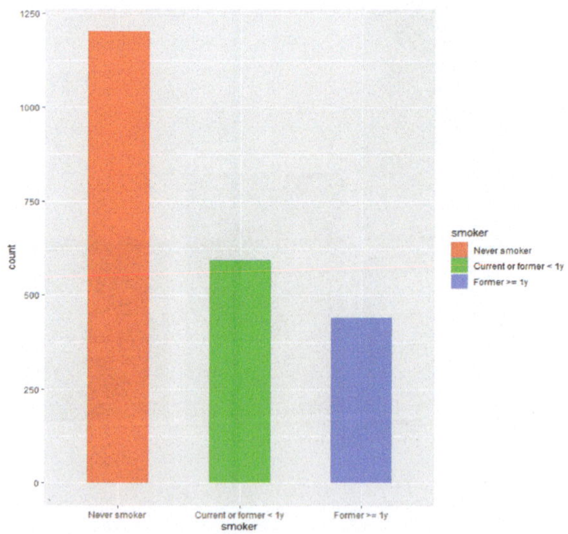

## 11.3.2 geom_bar() for Percentage

*# to get percentages*

```
prop.table(table(dat$smoker))
##
## Never smoker Current or former < 1y   Former >= 1y
##   0.5378415   0.2655620   0.1965965
ggplot(data = dat, aes(x= 1, fill = smoker))+geom_bar(position = "fill")+

  scale_y_continuous(name = "percentage",labels = scales::percent_format())
```

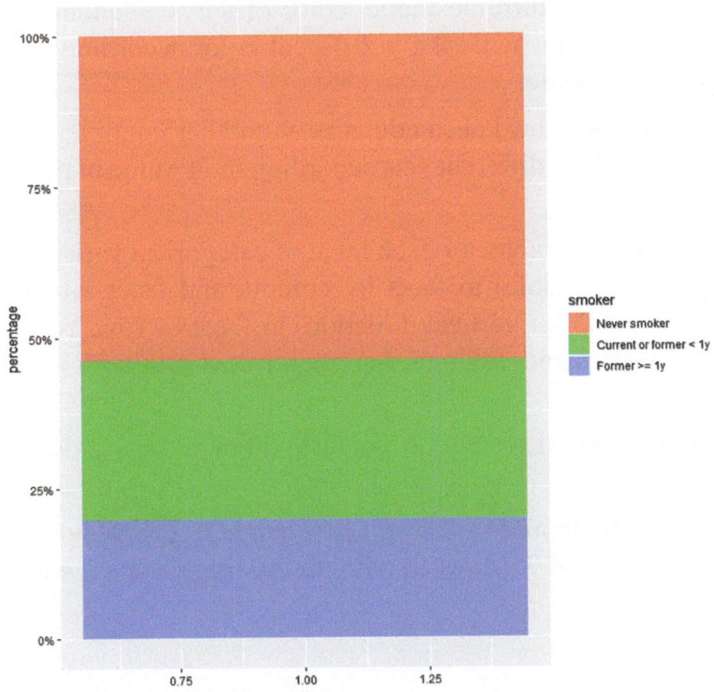

Notice how the bar heights are equal to counts from table() function and also the percentages in stacked bar plot are equal to that of prop.table(table()) function.

## 11.4 BIVARIATE ANALYSIS: CONTINUOUS-CONTINUOUS DATA

### 11.4.1 geom_point() for Two Variables

geom_point() is used to create a scatterplot. The scatterplot is most useful for displaying the relationship between two continuous variables. A bubble chart is a scatterplot with a third variable mapped to the size of points.

geom_point() requires two aesthetics (x, one of the continuous variables and y, the other continuous variable).

You can add another third aesthetic, color or shape = another categorical variable so every point will take a different color according to its level of this categorical variable.

You can add another third aesthetic, size = another continuous variable so every point will take a different size according to its value of this continuous variable.

To produce separate graphs for each level of categorical variable, use facet_wrap(categorical variable) to facet by column and facet_wrap(categorical variable~., dir = "v" for vertical direction) to facet by row. You can control the number of rows and columns for the facetted plot with nrow and ncol arguments.

Example, scatterplot of weight on x-axis and height on y-axis from regicor data.

**ggplot**(data = regicor, **aes**(x = weight, y = height))+**geom_point**()
## Warning: Removed 35 rows containing missing values (geom_point).

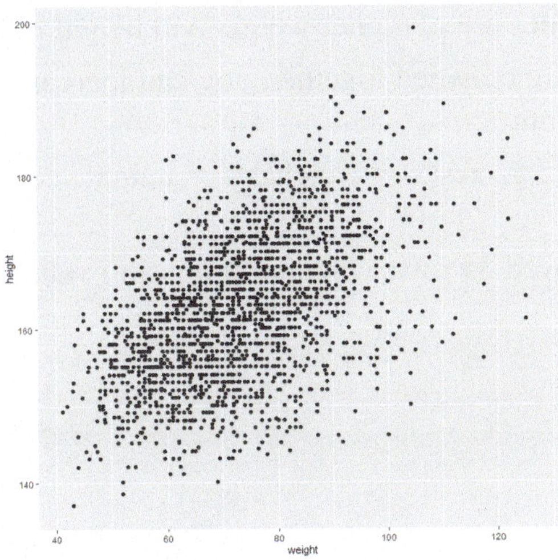


## 11.4.2 geom_point() for Three Variables

Add a third aesthetic, color = sex to color the points according to gender value.

**ggplot**(data = regicor, **aes**(x = weight, y = height, color = sex))+**geom_point**()

## Warning: Removed 35 rows containing missing values (geom_point).

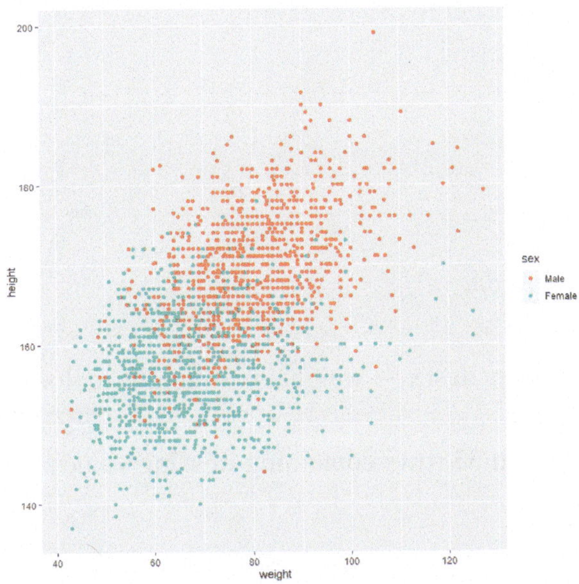

Notice how males are of higher weight and height than females.

As the points are crowded together, you can increase transparency with alpha to see all points.

**ggplot**(data = regicor, **aes**(x = weight, y = height, color = sex))+**geom_point**(alpha=0.5)

## Warning: Removed 35 rows containing missing values (geom_point).

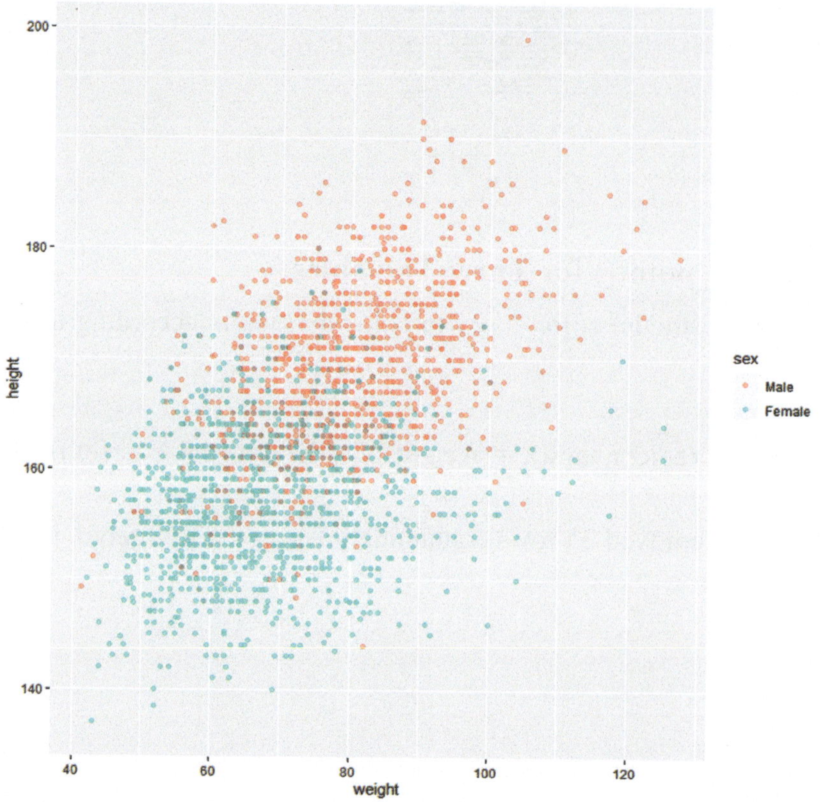


*# to be more transparent*

**ggplot**(data = regicor, **aes**(x = weight, y = height, color = sex))+**geom_point**(alpha=0.3)

## Warning: Removed 35 rows containing missing values (geom_point).

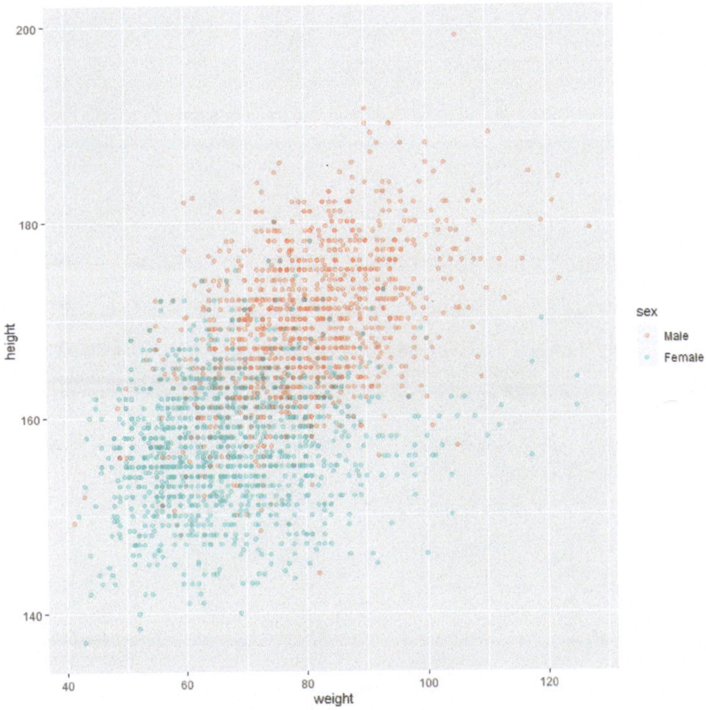


## 11.4.3 geom_point() for Four Variables

Add a fourth aesthetic, size = age to size the points according to the age value.

**ggplot**(data = regicor, **aes**(x = weight, y = height, color = sex, size = age))+

**geom_point**(alpha=0.3)

## Warning: Removed 35 rows containing missing values (geom_point).

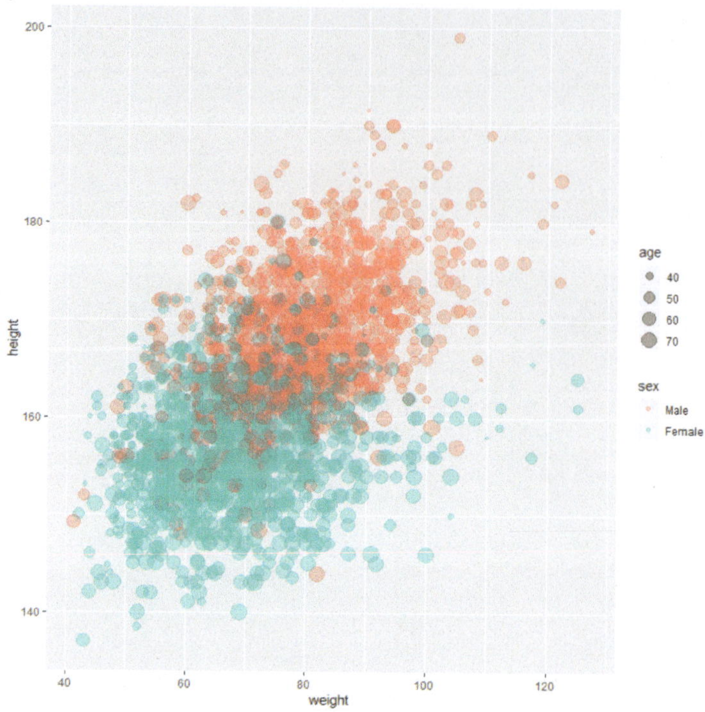


### 11.4.4 geom_point() for Five Variables

Add a fifth aesthetic, shape = smoker so the points will get different shapes according to the smoker status.

**ggplot**(data = regicor, **aes**(x = weight, y = height, color = sex, shape = smoker, size = age))+

**geom_point**(alpha=0.5)

## Warning: Removed 85 rows containing missing values (geom_point).

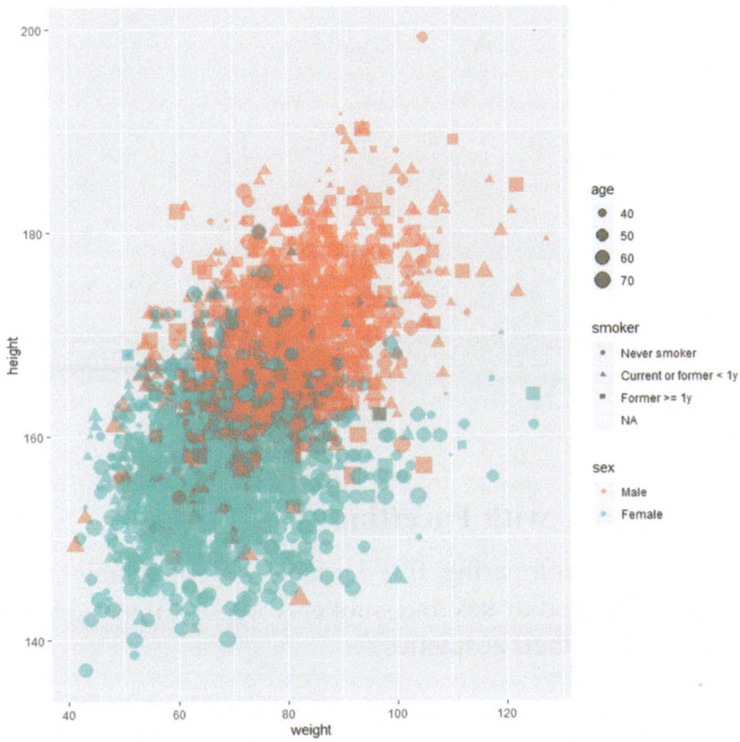


*# to get rid of NA values*

dat<-**filter**(regicor, **!is.na**(smoker))

**ggplot**(data = dat, **aes**(x = weight, y = height, color = sex, shape = smoker, size = age))+

**geom_point**(alpha=0.5)

## Warning: Removed 24 rows containing missing values (geom_point).

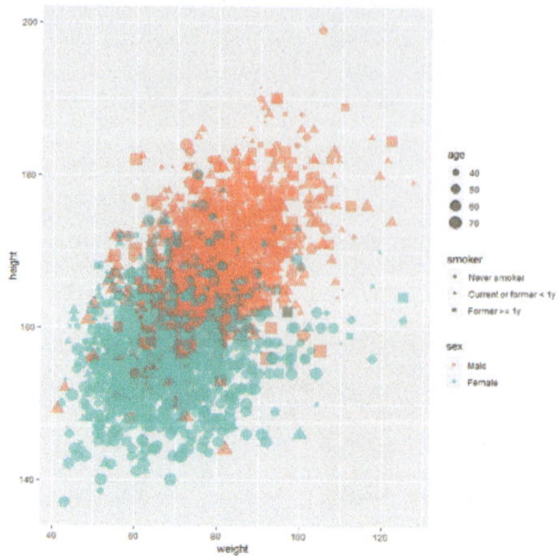


## 11.4.5 geom_point() with Facetting

As the last graph is comparing five variables, weight, height, age, sex, smoker, it can be facetted by sex and smoker (categorical variables) and you do not need to specify their aesthetics.

*#facet by sex by column*

**ggplot**(data = dat, **aes**(x = weight, y = height, shape = smoker, size = age))+

**geom_point**(alpha=0.5)+**facet_wrap**(.~sex)

## Warning: Removed 24 rows containing missing values (geom_point).

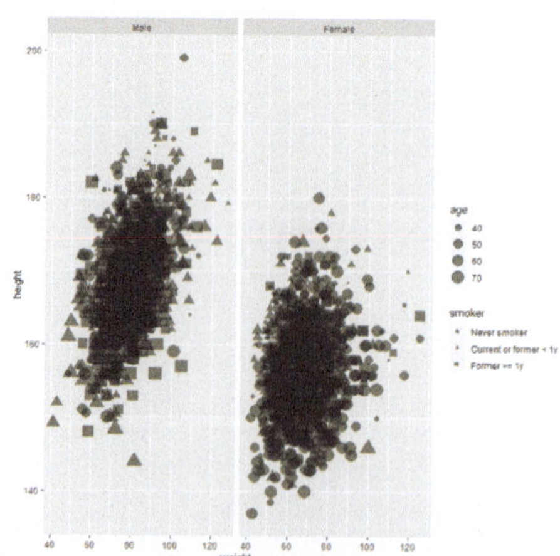

# facet by sex by row

**ggplot**(data = dat, **aes**(x = weight, y = height, shape = smoker, size = age))+

**geom_point**(alpha=0.5)+**facet_wrap**(sex~., dir = "v")

## Warning: Removed 24 rows containing missing values (geom_point).

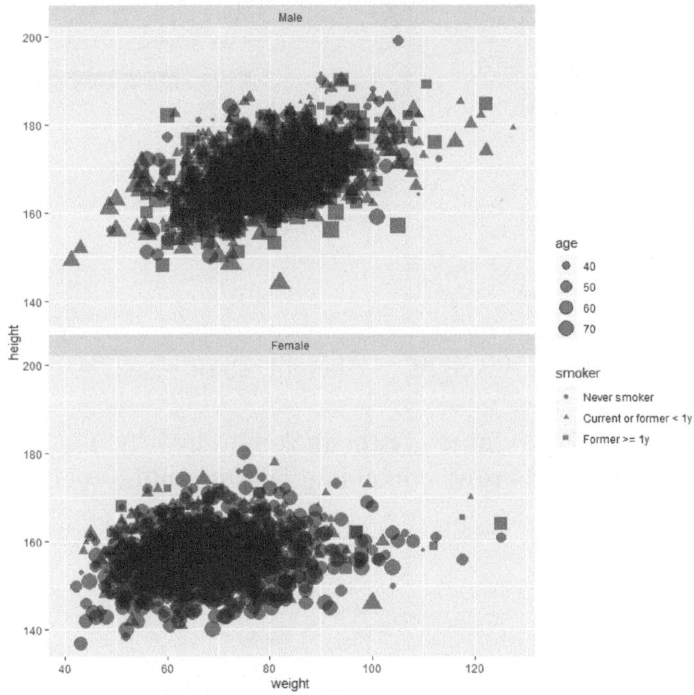


*# facet by smoker by column*

**ggplot**(data = dat, **aes**(x = weight, y = height, color = sex, size = age))+

**geom_point**(alpha=0.5)+**facet_wrap**(.~smoker)

## Warning: Removed 24 rows containing missing values (geom_point).

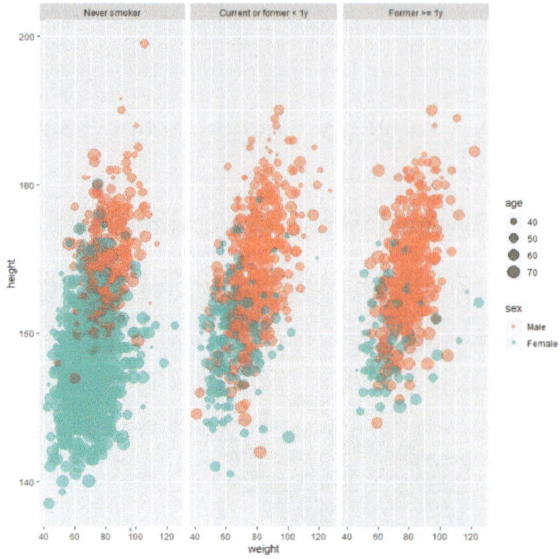


*# facet by smoker by row*

**ggplot**(data = dat, **aes**(x = weight, y = height, color = sex, size = age))+

**geom_point**(alpha=0.5)+**facet_wrap**(smoker~., dir = “v”)
## Warning: Removed 24 rows containing missing values (geom_point).

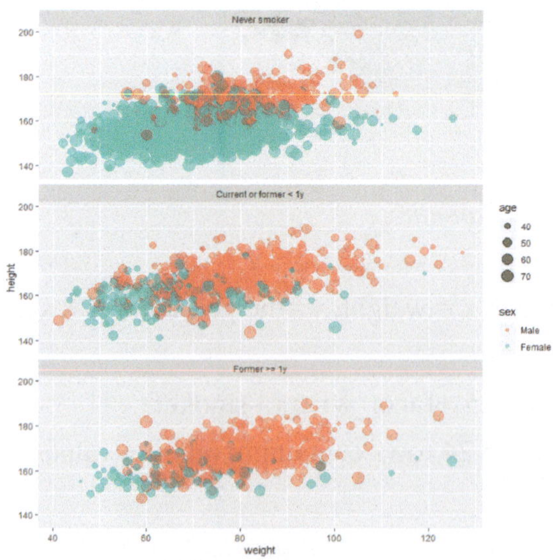

*# facet by smoker as row and sex as column and add nrow = 3 = number of smoker levels*

**ggplot**(data = dat, **aes**(x = weight, y = height, size = age))+

**geom_point**(alpha=0.5)+**facet_wrap**(smoker~sex, nrow = 3)
## Warning: Removed 24 rows containing missing values (geom_point).

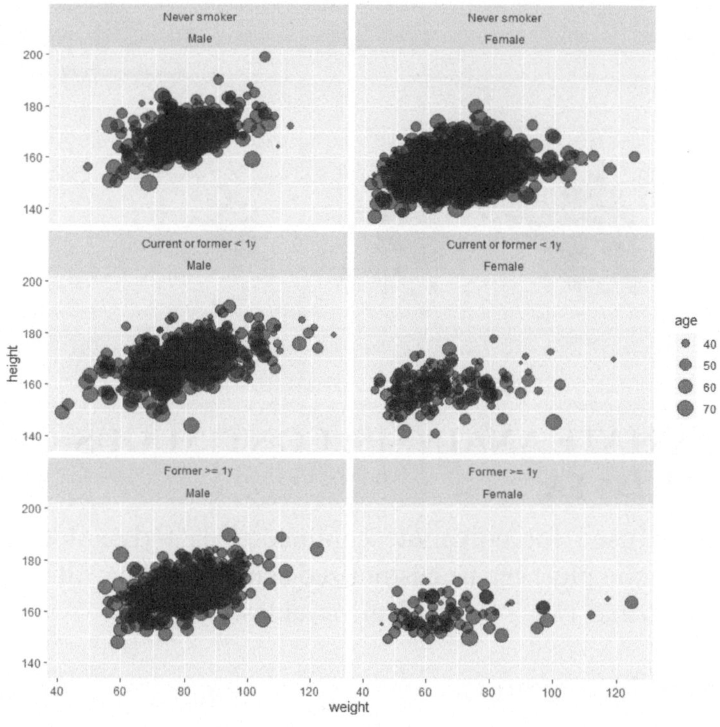


*# facet by smoker as column and sex as row and add nrow = 2 = number of sex levels*

**ggplot**(data = dat, **aes**(x = weight, y = height, size = age))+

**geom_point**(alpha=0.5)+**facet_wrap**(sex ~ smoker, nrow = 2)
## Warning: Removed 24 rows containing missing values (geom_point).

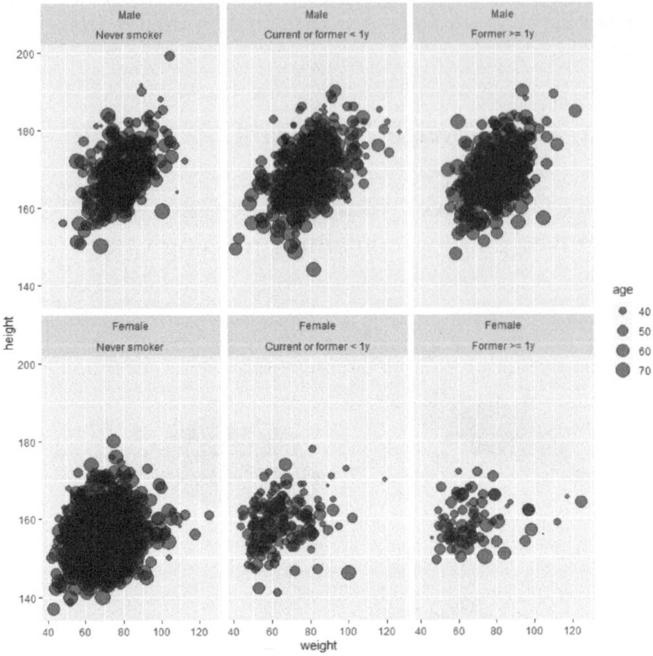


## 11.5 BIVARIATE ANALYSIS: CONTINUOUS-CATEGORICAL DATA

All plot types used for univariate continuous data (boxplot, histogram, density plot, violin plot) can be customized using the color, fill aesthetics, or facet_wrap() to add 1 or 2 categorical variables.

### 11.5.1 geom_histogram()

Example, histogram of regicor data weight by smoker or by smoker and sex.

```
# remove NA values

dat<-filter(regicor, !is.na(smoker))
```

*# by smoker*

**ggplot**(data = dat, **aes**(x = weight, fill = smoker))+**geom_histogram**()
## `stat_bin()` using `bins = 30`. Pick better value with `binwidth`.
## Warning: Removed 24 rows containing non-finite values (stat_bin).

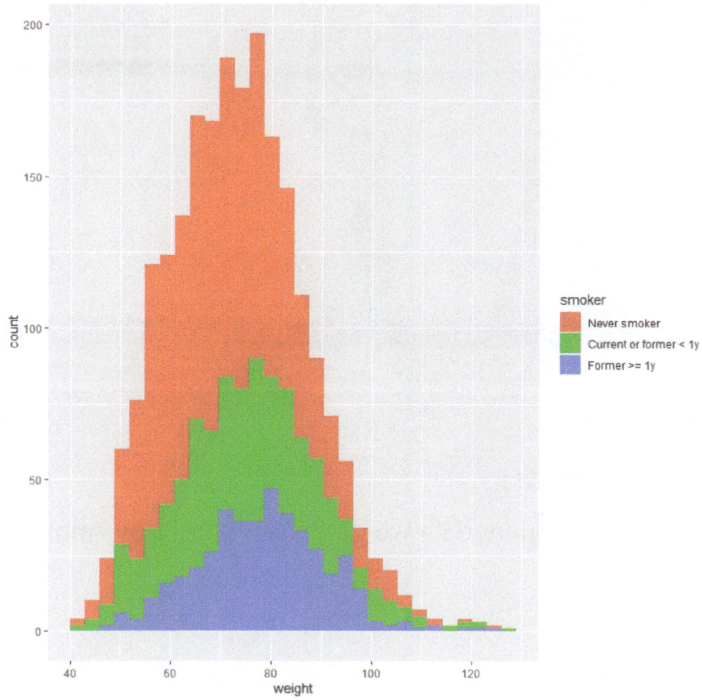


*# by smoker and sex*

**ggplot**(data=dat, **aes**(x=weight, fill=smoker))+**geom_histogram**()+**facet_wrap**(~sex)
## `stat_bin()` using `bins = 30`. Pick better value with `binwidth`.
## Warning: Removed 24 rows containing non-finite values (stat_bin).

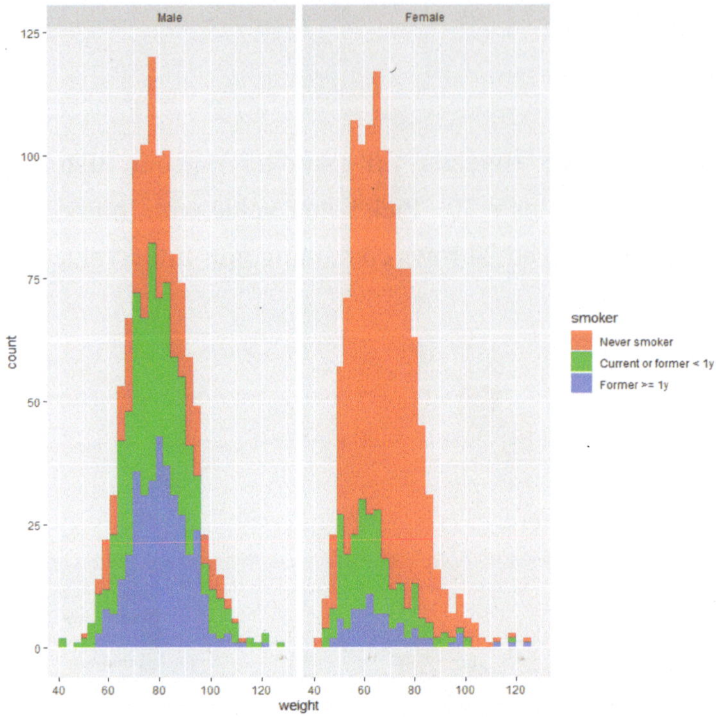


## 11.5.2 geom_boxplot()

Example, boxplot of regicor data weight by smoker or by smoker and sex.

```
# remove NA values

dat<-filter(regicor, !is.na(smoker))

# by smoker

ggplot(data = dat, aes(x = smoker, y = weight, fill = smoker))+geom_
boxplot()
## Warning: Removed 24 rows containing non-finite values (stat_boxplot).
```

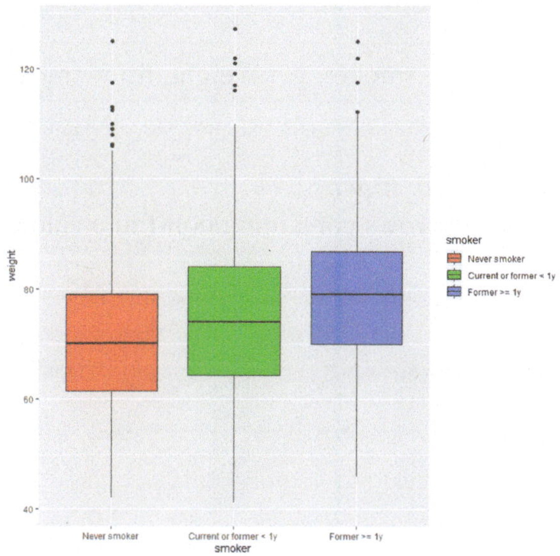


*# by smoker and sex*

**ggplot**(data = dat, **aes**(x = smoker, y = weight, fill = smoker))+**geom_boxplot**()+**facet_wrap**(~sex)

## Warning: Removed 24 rows containing non-finite values (stat_boxplot).

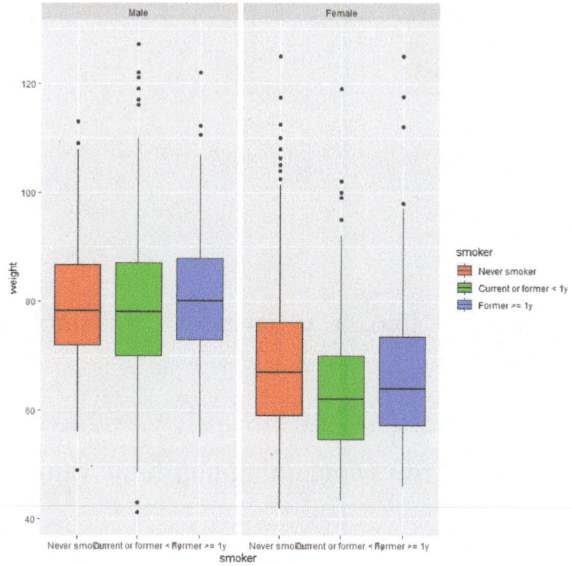

*# many boxplots so use coord_flip()*

**ggplot**(data = dat, **aes**(x = smoker, y = weight, fill = smoker))+**geom_boxplot**()+

**facet_wrap**(~sex)+ **coord_flip**()
## Warning: Removed 24 rows containing non-finite values (stat_boxplot).

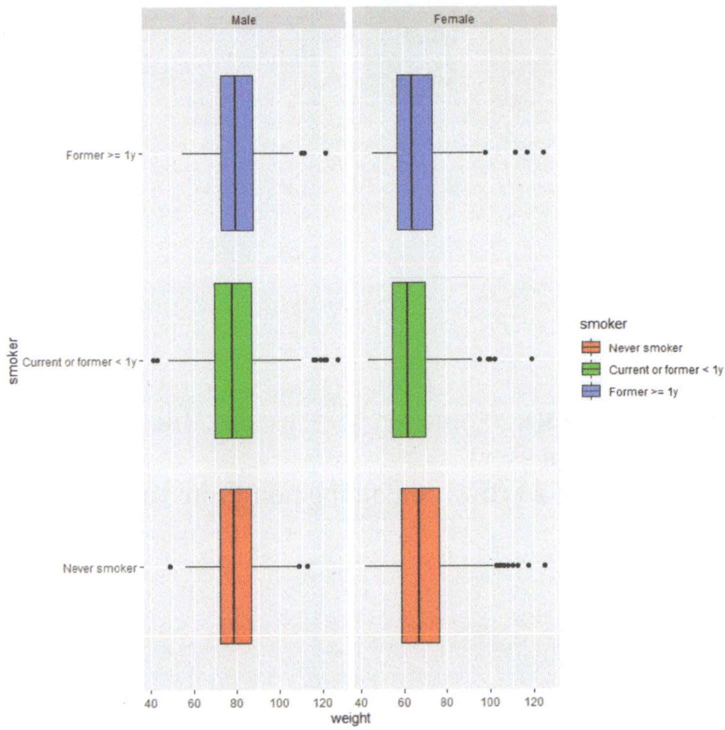


*# or facet by row*

**ggplot**(data = dat, **aes**(x = smoker, y = weight, fill = smoker))+**geom_boxplot**()+

**facet_wrap**(sex~., dir = "v")
## Warning: Removed 24 rows containing non-finite values (stat_boxplot).

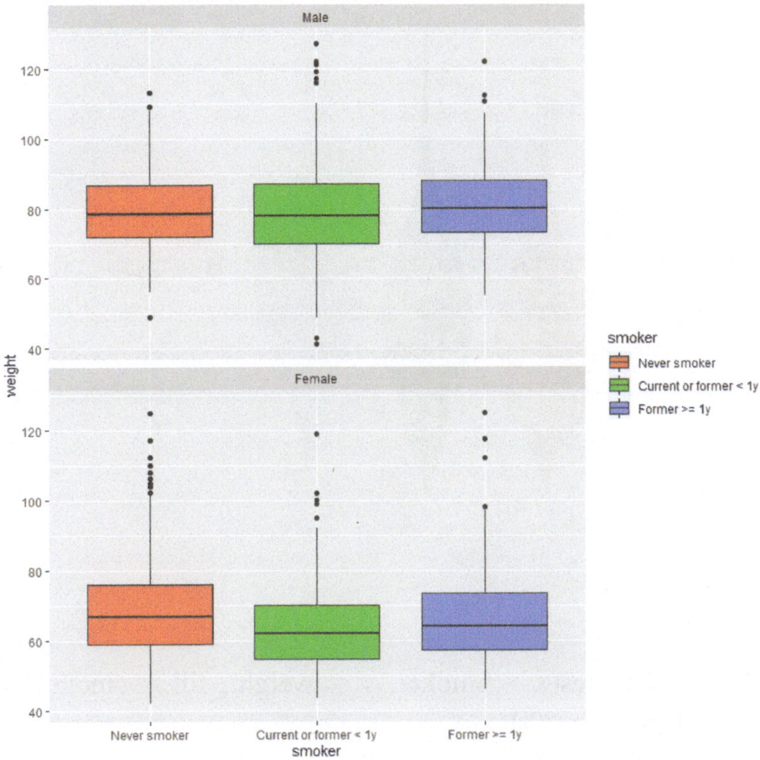


## 11.5.3 geom_violin()

Example, violin plot of regicor data weight by smoker or by smoker and sex.

```
# remove NA values

dat<-filter(regicor, !is.na(smoker))

# by smoker

ggplot(data = dat, aes(x = smoker, y = weight, fill = smoker))+geom_violin()
## Warning: Removed 24 rows containing non-finite values (stat_ydensity).
```

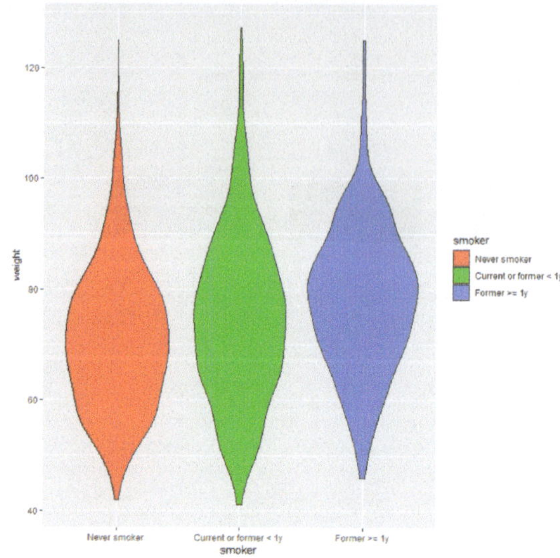


*# by smoker and sex*

```
ggplot(data = dat, aes(x = smoker, y = weight, fill = smoker))+geom_violin()+facet_wrap(~sex)
## Warning: Removed 24 rows containing non-finite values (stat_ydensity).
```

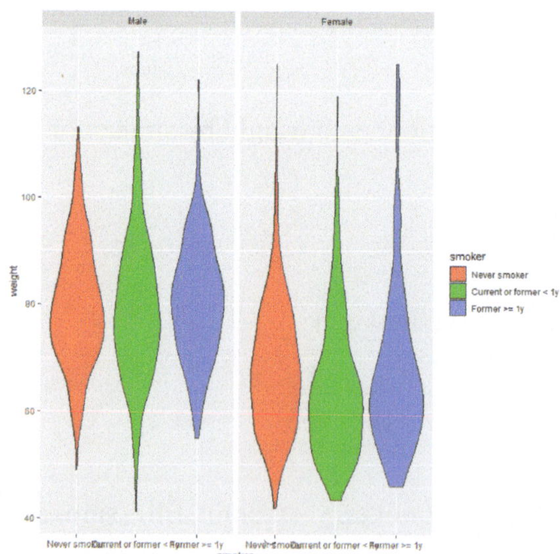


# many violins so use coord_flip()

**ggplot**(data = dat, **aes**(x = smoker, y = weight, fill = smoker))+**geom_violin**()+

**facet_wrap**(~sex)+ **coord_flip**()

## Warning: Removed 24 rows containing non-finite values (stat_ydensity).

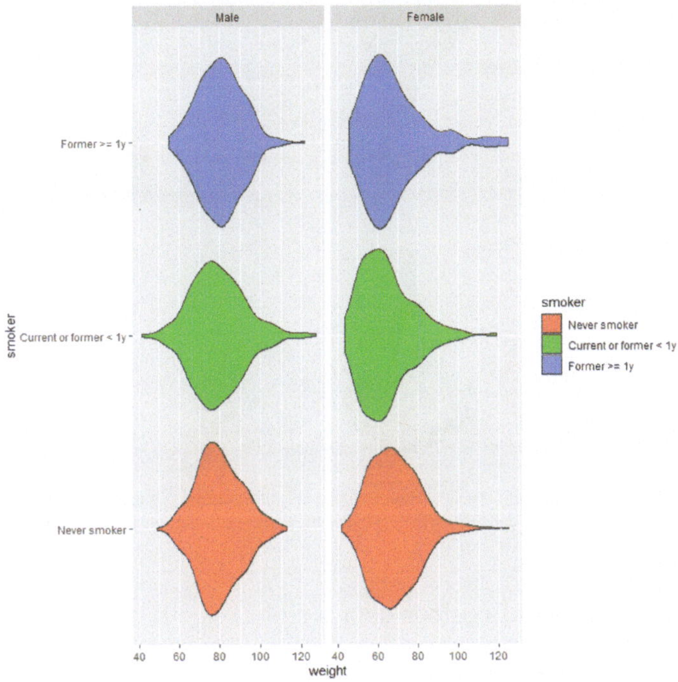


# or facet by row

**ggplot**(data = dat, **aes**(x = smoker, y = weight, fill = smoker))+**geom_violin**()+

**facet_wrap**(sex~., dir = "v")

## Warning: Removed 24 rows containing non-finite values (stat_ydensity).

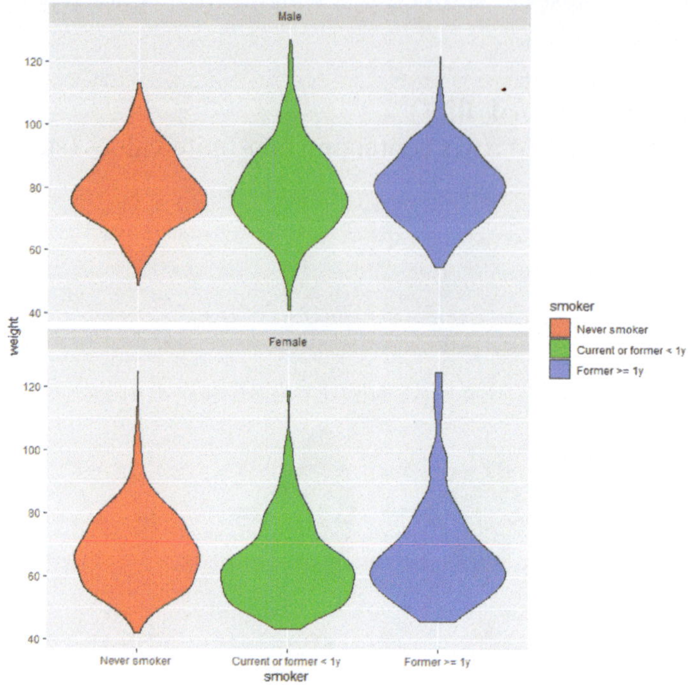


### 11.5.4 geom_density()

Example, density plots of regicor data weight by smoker or by smoker and sex.

```
# remove NA values

dat<-filter(regicor, !is.na(smoker))

# by smoker

ggplot(data = dat, aes(x = weight, fill = smoker))+geom_density()
## Warning: Removed 24 rows containing non-finite values (stat_density).
```

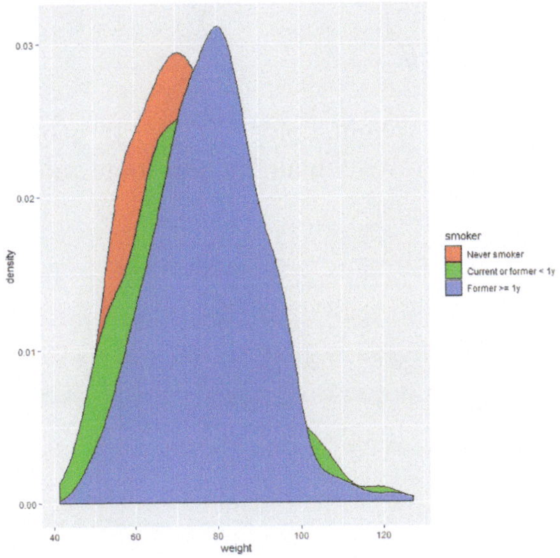


*# to see 3 curves, add alpha*

**ggplot**(data = dat, **aes**(x = weight, fill = smoker))+**geom_density**(alpha = 0.5)
## Warning: Removed 24 rows containing non-finite values (stat_density).

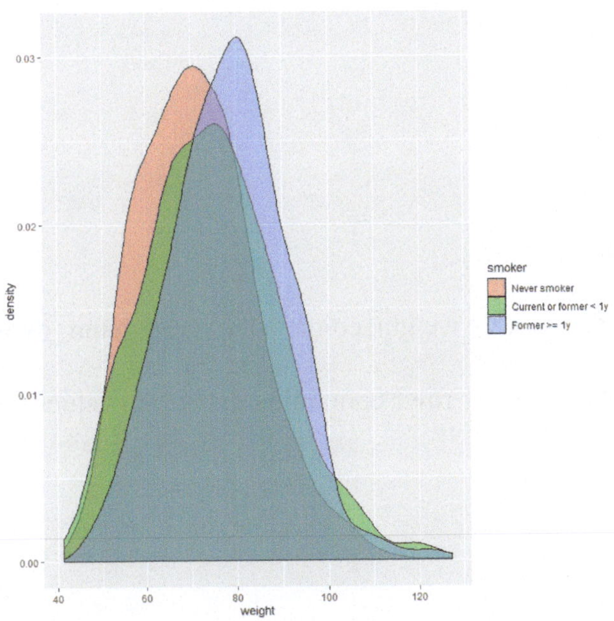

*# or add color instead of fill*

**ggplot**(data = dat, **aes**(x = weight, color = smoker))+**geom_density**()
## Warning: Removed 24 rows containing non-finite values (stat_density).

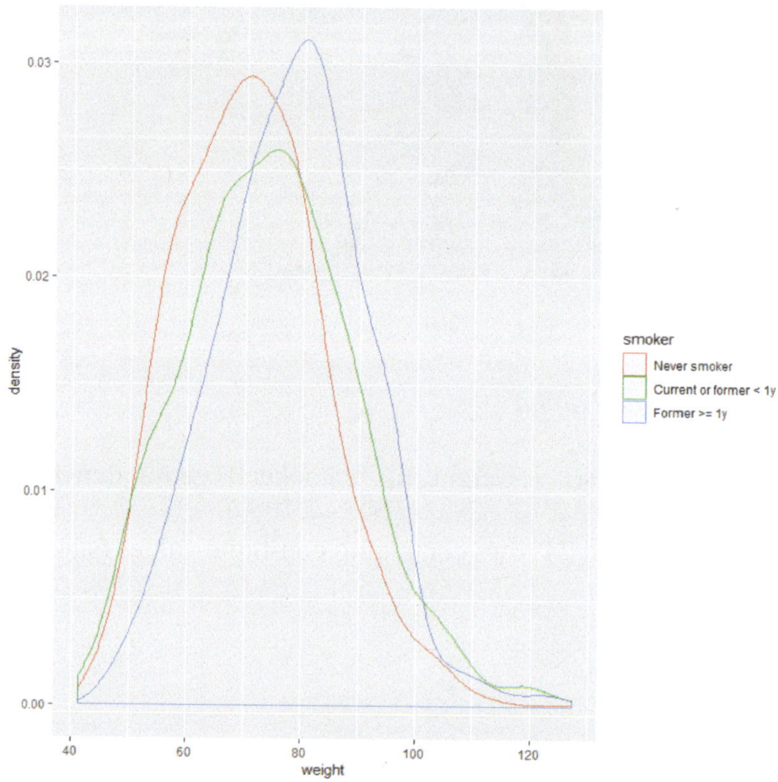


*# by smoker and sex*

**ggplot**(data = dat, **aes**(x = weight, color = smoker))+**geom_density**()+**facet_wrap**(~sex)
## Warning: Removed 24 rows containing non-finite values (stat_density).

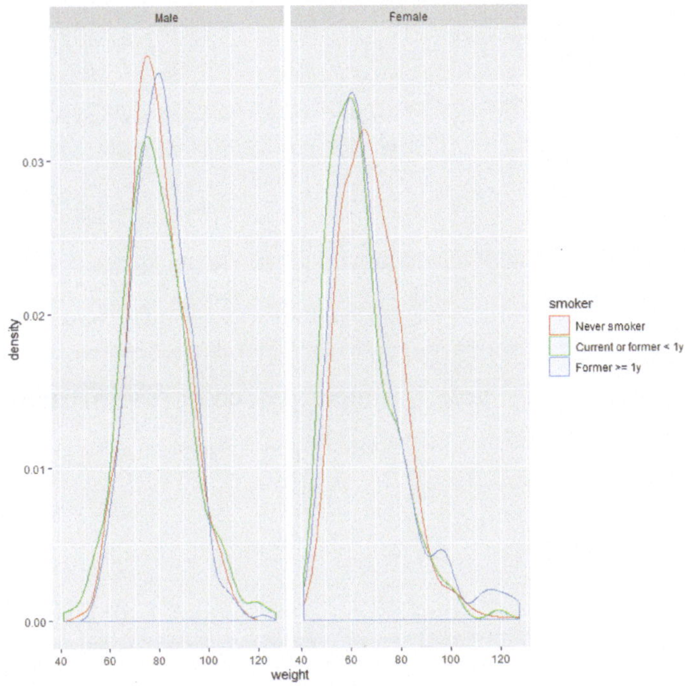


## 11.5.5 One Continuous (Summary) Value for Each Category

In other case, if you have a factor and one value for each factor, you can use geom_point() or geom_bar (with stat = "identity", position = "dodge")

geom_point() is used for time-series data, for example, the mean weight of each sex in each year (1995,2000,2005). To make the plot clearer, we add another aesthetic, group = sex, then add geom_line() to connect points of the same group.

```
dat<-regicor %>% group_by(year, sex) %>% summarise(mean_wt = mean(weight, na.rm = TRUE))

dat
## # A tibble: 6 x 3
## # Groups: year [3]
## year sex mean_wt
## <fct> <fct> <dbl>
```

```
## 1 1995 Male  77.6
## 2 1995 Female 67.3
## 3 2000 Male  80.1
## 4 2000 Female 67.6
## 5 2005 Male  80.2
## 6 2005 Female 67.7
```

*# plot the data*

**ggplot**(data = dat, **aes**(x = year, y = mean_wt, color = sex))+**geom_point**()

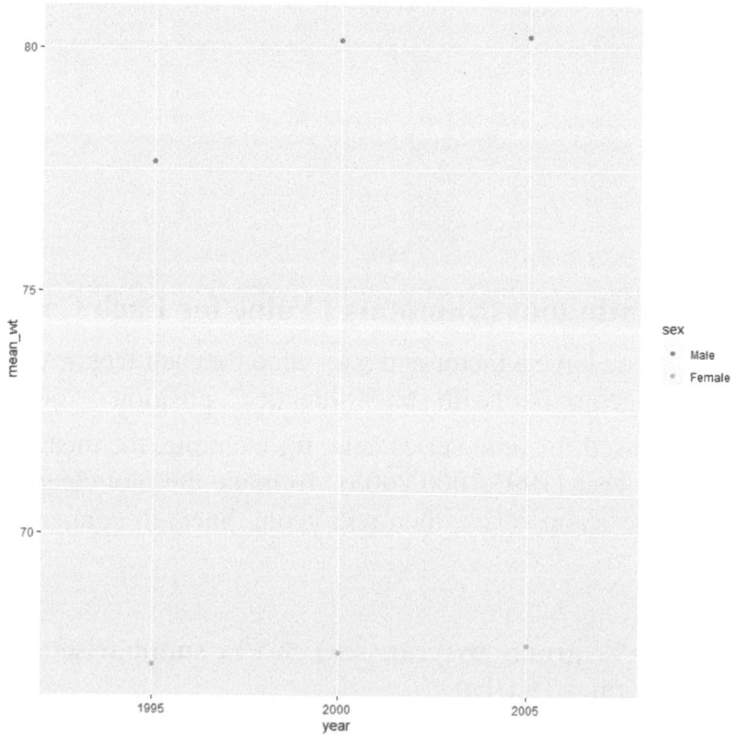


*# Add lines*

**ggplot**(data = dat, **aes**(x = year, y = mean_wt, color = sex, group = sex))+

**geom_point**()+**geom_line**()

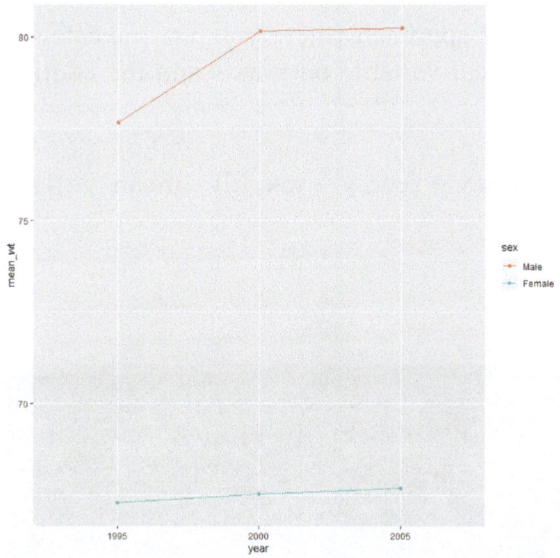


# plot bar graphs

**ggplot**(data = dat, **aes**(x = year, y = mean_wt, fill = sex))+

**geom_bar**(stat = "identity", position = "dodge")

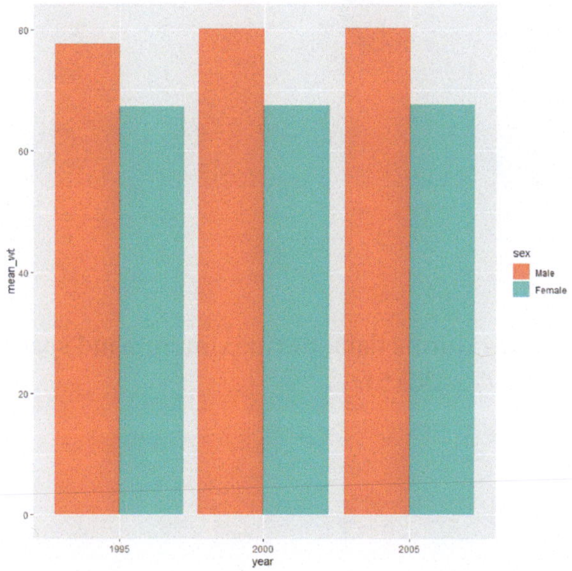

### *11.5.6. geom_tile()*

geom_tile() is another approach where you put one categorical variable on x-axis. other categorical variable on y-axis and the continuous variable to color the cells.

**ggplot**(data = dat, **aes**(x = year, y = sex, fill = mean_wt))+

**geom_tile**()

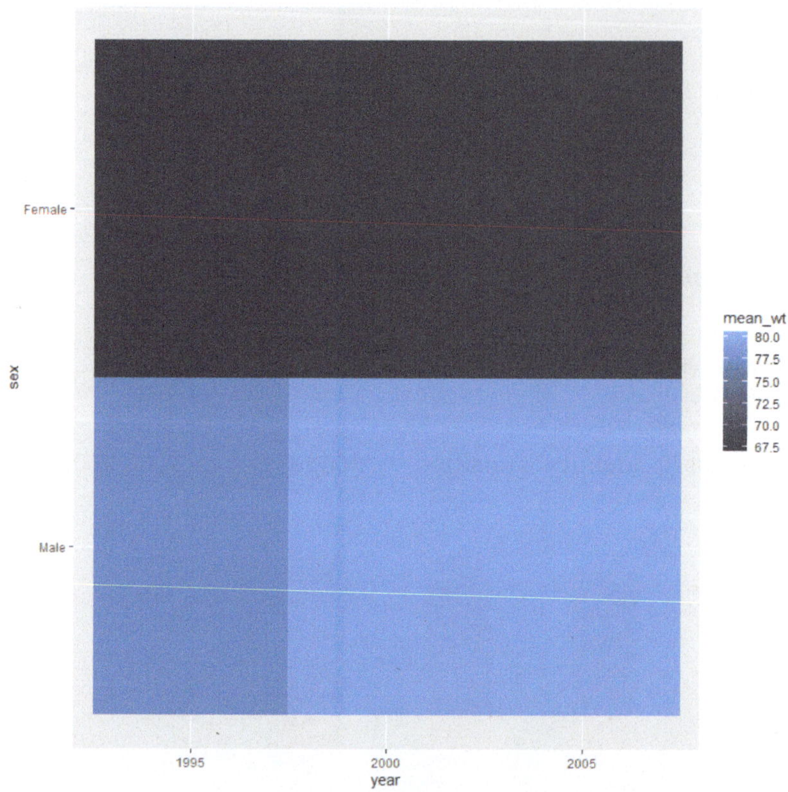


Another example from storms data, the maximum wind speed for each wind status in each year (1975–2015).

dat<-storms **%>%** **group_by**(year, status) **%>%** **summarise**(max_wind = **max**(wind, na.rm = TRUE))

*# plot the data*

**ggplot**(data = dat, **aes**(x = year, y = max_wind, color = status))+**geom_point**()

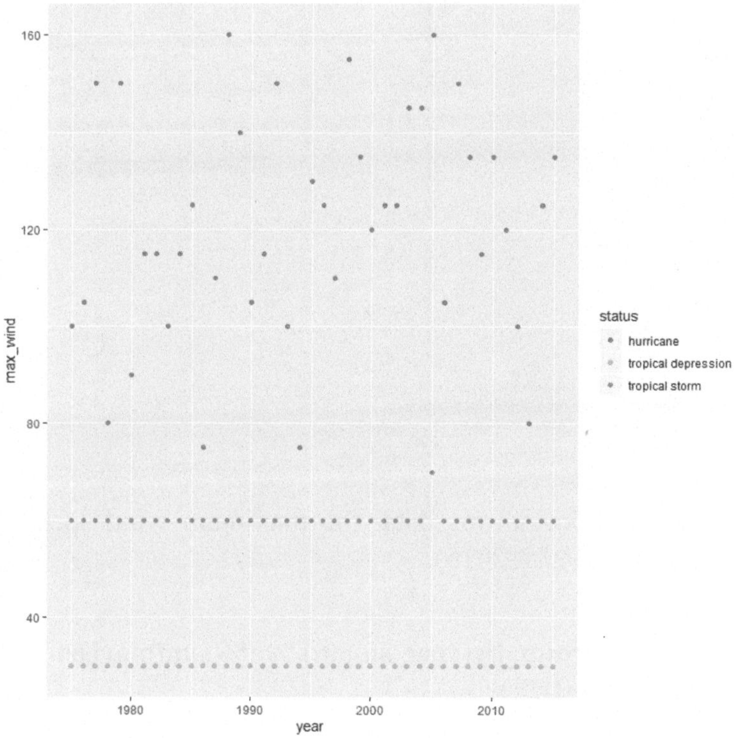


*# Add lines*

**ggplot**(data = dat, **aes**(x = year, y = max_wind, color = status, group = status))+

**geom_point**()+**geom_line**()

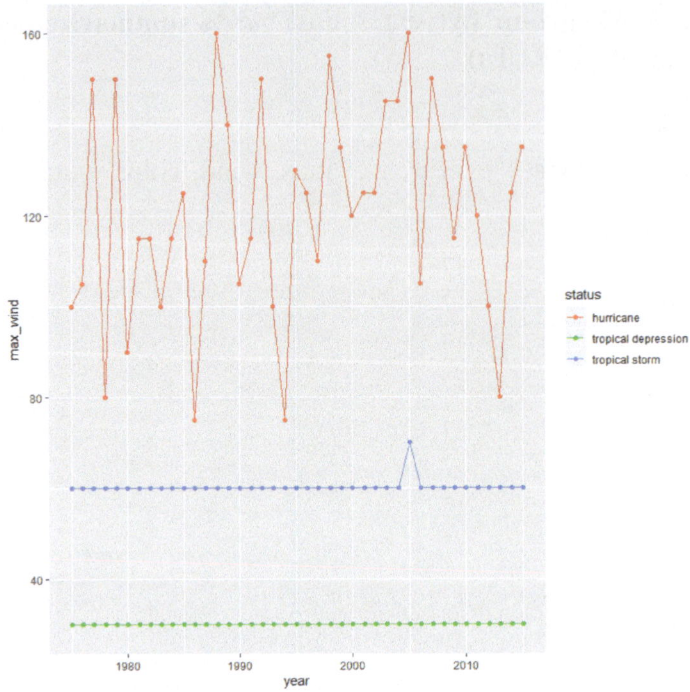


Another example from storms data, the maximum wind speed for each month in each year (1975–2015).

```
dat<-storms %>% group_by(year, month) %>% summarise(max_wind = max(wind, na.rm = TRUE))

# covert month to a factor
dat$month<-factor(dat$month)

# plot the data

ggplot(data = dat, aes(x = year, y = max_wind, color = month))+geom_point()
```

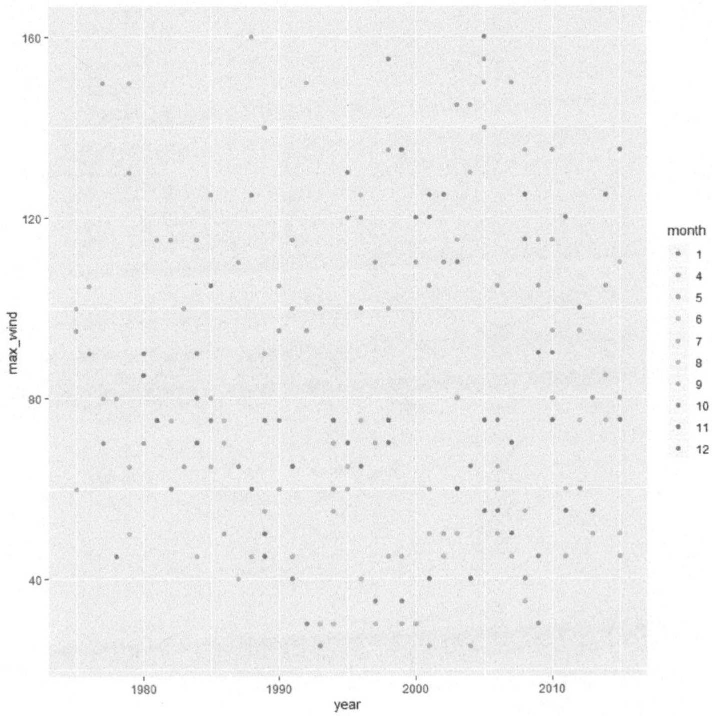


*# Add lines*

**ggplot**(data = dat, **aes**(x = year, y = max_wind, color = month,group = month))+

**geom_point()+geom_line()**

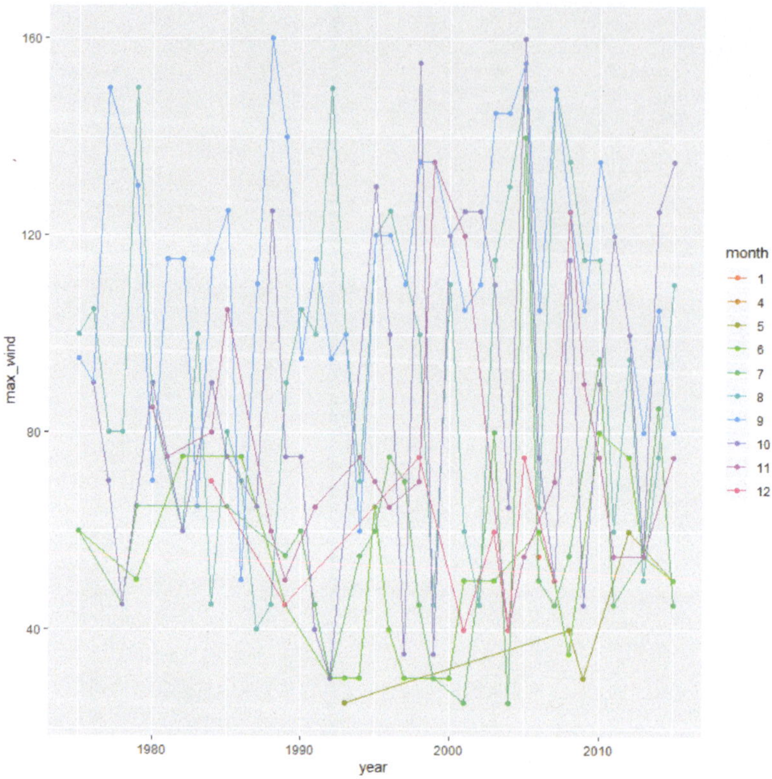


*#many lines so facet by month*

**ggplot**(data = dat, **aes**(x = year, y = max_wind))+

**geom_point**()+**geom_line**()+**facet_wrap**(~month, nrow = 4)

## geom_path: Each group consists of only one observation. Do you need to adjust

## the group aesthetic?

## geom_path: Each group consists of only one observation. Do you need to adjust

## the group aesthetic?

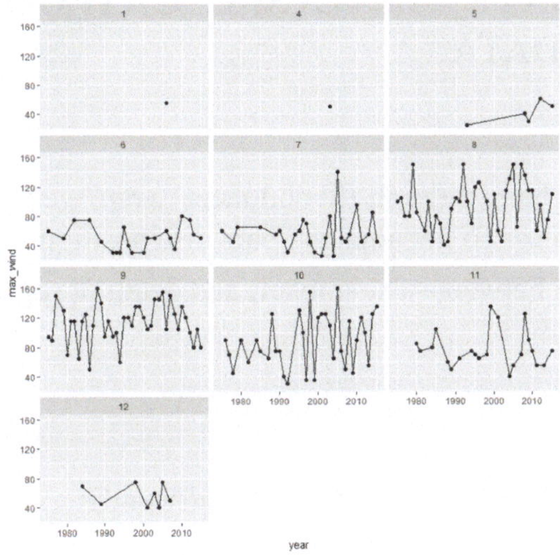


*# geom_tile*

**ggplot**(data = dat, **aes**(x = month, y = year, fill= max_wind))+

**geom_tile**()

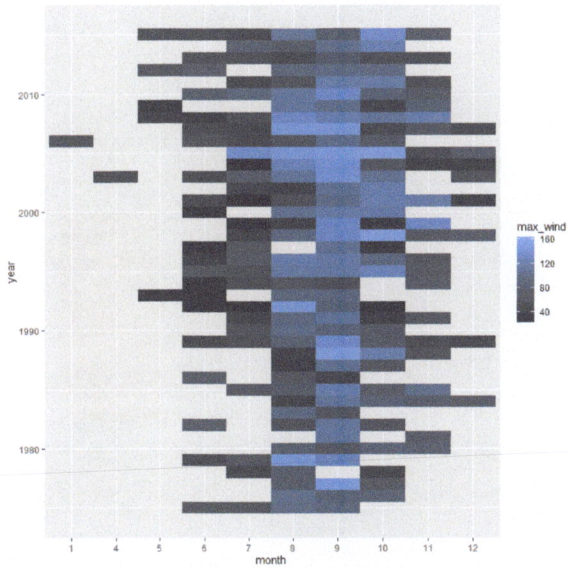

## 11.6 BIVARIATE ANALYSIS: CATEGORICAL-CATEGORICAL DATA

### 11.6.1 Counts

for counts, geom_bar() through fill and facet_wrap, we can add another two categorical variables. To get adjacent bars, use position = "dodge"

Example, relation between sex, smoker, and death of regicor data.

```
# filter for NA

dat<-filter(regicor, !is.na(smoker) & !is.na(death))

ggplot(data = dat, aes(x= smoker, fill = sex))+geom_bar(width = 0.5)
```

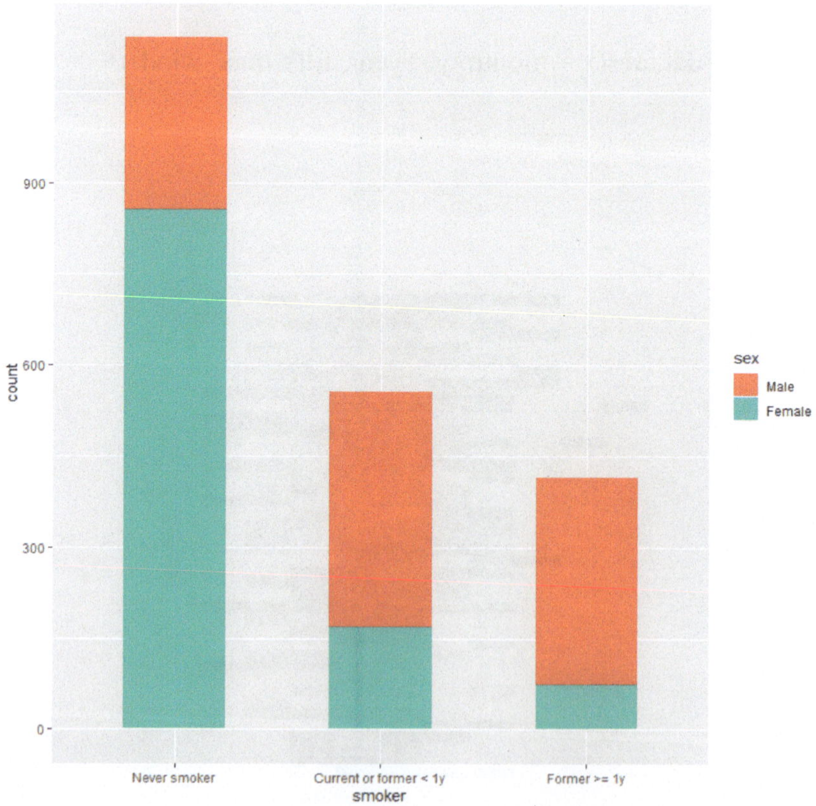

*# use position = "dodge"*

**ggplot**(data = dat, **aes**(x= smoker, fill = sex))+**geom_bar**(width = 0.5, position = "dodge")

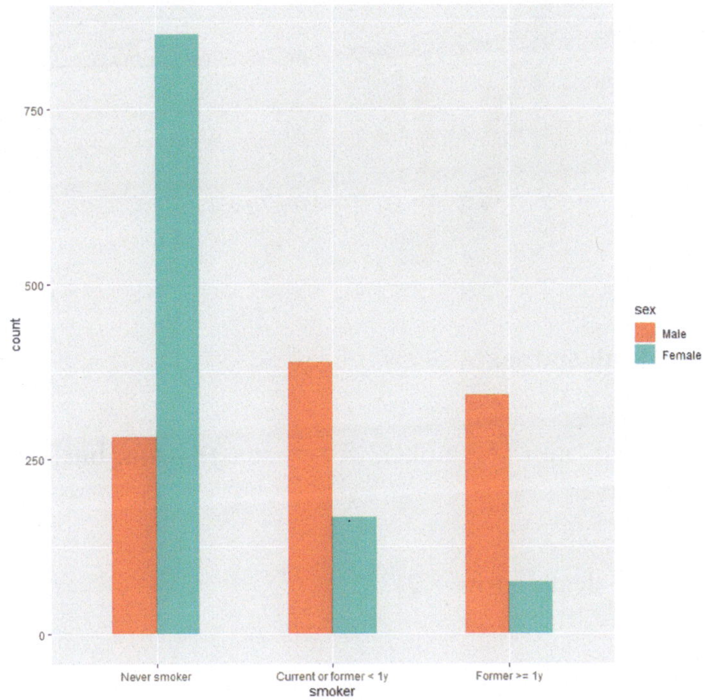


*# To facet by death*

**ggplot**(data = dat, **aes**(x= smoker, fill = sex))+**geom_bar**(width = 0.5, position = "dodge")+

**facet_wrap**(.~death)

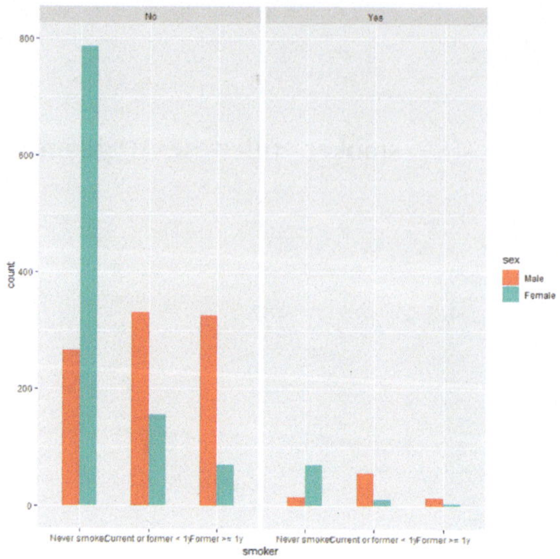


# To facet by death and sex

**ggplot**(data = dat, **aes**(x= smoker, fill = sex))+**geom_bar**(width = 0.5, position = "dodge")+

**facet_wrap**(sex~death, nrow = 2)

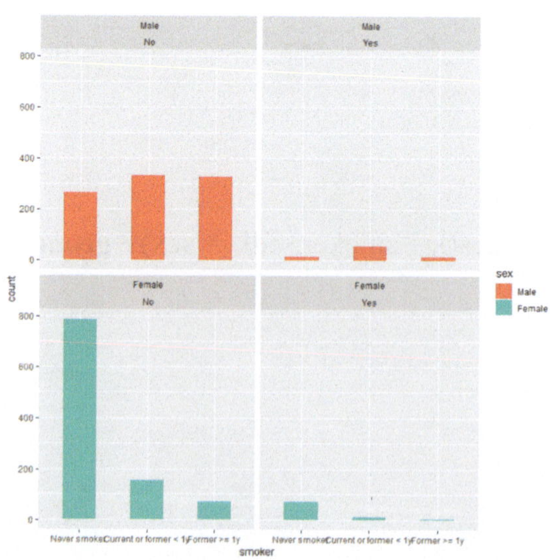

*# many labels so use coord_flip()*

**ggplot**(data = dat, **aes**(x= smoker, fill = sex))+**geom_bar**(width = 0.5, position = "dodge")+
**facet_wrap**(sex~death, nrow = 2) + **coord_flip**()

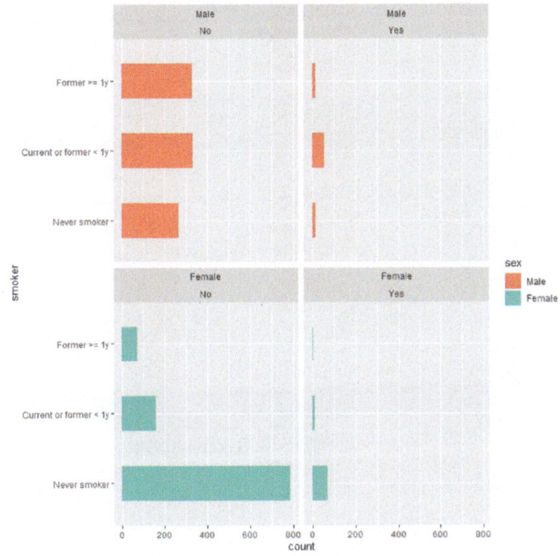


## 11.6.2 Percentages

For percentages, geom_bar() through fill and facet_wrap, we can add another two categorical variables. You must use position = "fill".

Example, relation between sex, smoker, and death.

*# filter for NA*

dat<-**filter**(regicor, **!is.na**(smoker) **& !is.na**(death))

*# use position = "dodge"*

**ggplot**(data = dat, **aes**(x= smoker, fill = sex))+**geom_bar**(width = 0.5, position = "fill")+

**scale_y_continuous**(name = "percentage",labels = scales::**percent_format**())

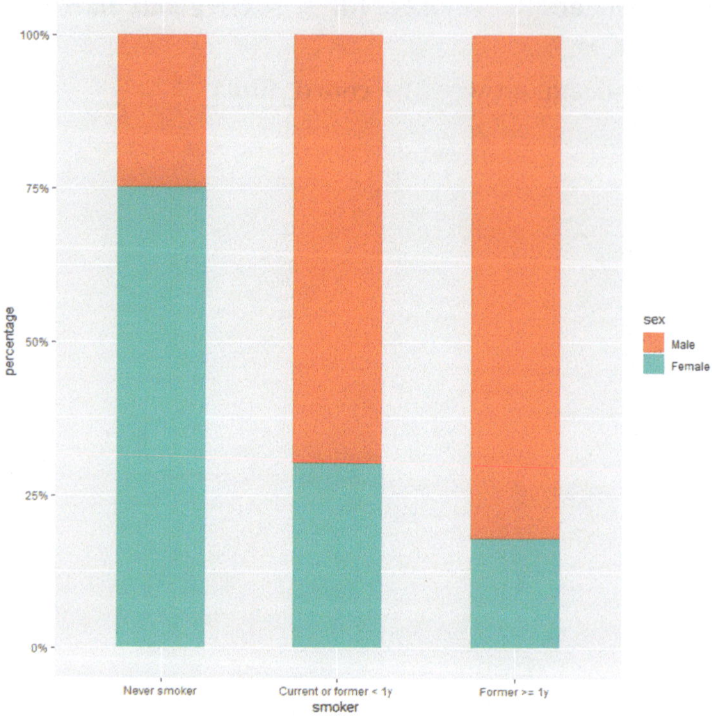


*# To facet by death*

**ggplot**(data = dat, **aes**(x= smoker, fill = sex))+**geom_bar**(width = 0.5, position = "fill")+

**scale_y_continuous**(name = "percentage",labels = scales::**percent_format**())+

**facet_wrap**(.~death)

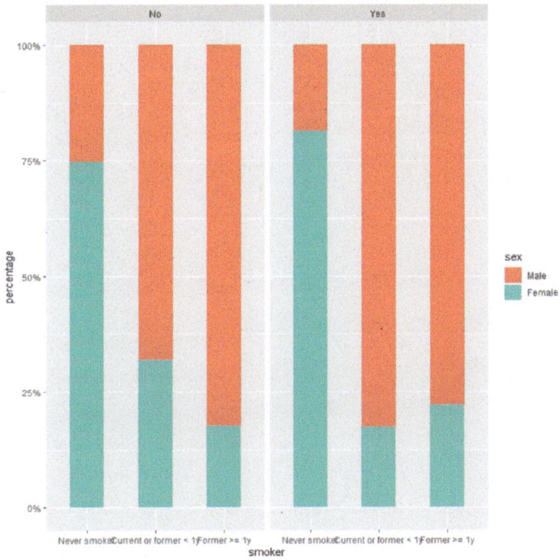


*# many labels so use coord_flip()*

**ggplot**(data = dat, **aes**(x= smoker, fill = sex))+**geom_bar**(width = 0.5, position = "fill")+

**scale_y_continuous**(name = "percentage",labels = scales::**percent_format**())+

**facet_wrap**(.~death) + **coord_flip**()

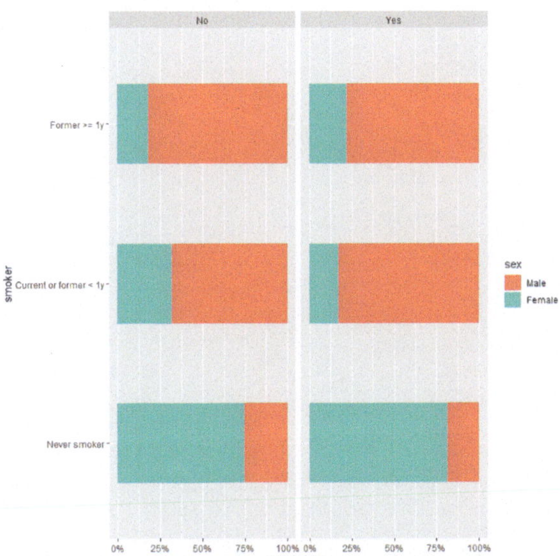

# CHAPTER 12

# FUNCTIONS

## CONTENTS

## 12.1 FUNCTIONS

Functions are small pieces of reusable codes. They are characterized by the function name followed by parentheses. The parentheses contain arguments that the function operates on.

Example, the mean() function.

```
mean(c(2,3,6))
## [1] 3.666667
```

The mean() function returns the mean of the three numbers (2,3,6) given to it.

### 12.1.1 Creating a Function

When you create your own function with function(), the function will returns the last R expression within its curly braces {}. The following is the code syntax:

function_name<-function(argument){maniplulate arguments1 manipulate arguments2}

So the result of manipulate arguments2 will be returned by this function.

Example, create a function called calculate_mean that calculates the mean of a given vector.

```
calculate_mean<-function(x){

sum_x<-sum(x, na.rm = TRUE)

length_x<-length(x)-sum(is.na(x))

mean_x<-sum_x/length_x

mean_x

}
```

**calculate_mean**(1:10)

## [1] 5.5

**calculate_mean**(1:100)

## [1] 50.5

**calculate_mean**(12:555)

## [1] 283.5

**calculate_mean**(airquality$Ozone)

## [1] 42.12931

**mean**(airquality$Ozone, na.rm = TRUE)

## [1] 42.12931

Here we created the calculate_mean() function by the following steps:

1. We specify one argument called x;
2. Inside the curly braces, we calculate the sum of x (after removing NA values) and assign it to a variable called sum_x;
3. Then, we calculate the length of x (subtract from it the number of NA values in x) and assign it to a variable called length_x;
4. To calculate the mean_x, we divide sum_x/length_x;
5. The last expression written is mean_x so it is returned by our function.

As evident from the mean() and calculate_mean() functions, they return the same value for mean of its argument except that we do not need to specify the na.rm = TRUE argument for the new function, calculate_mean(), because it has already manipulated that inside its curly braces.

### 12.1.2 Function with Default Arguments

Functions with default arguments that are the most commonly used arguments but these arguments can be changed.

Example, the cut() function has a default behavior of not including the lowest value of the intervals but you can change this behavior.

x<–1:20

```
x
## [1] 1 2 3 4 5 6 7 8 9 10 11 12 13 14 15 16 17 18 19 20
y<-cut(x, breaks = c(1,10,20))

y
## [1] <NA> (1,10] (1,10] (1,10] (1,10] (1,10] (1,10] (1,10] (1,10]
## [10] (1,10] (10,20] (10,20] (10,20] (10,20] (10,20] (10,20] (10,20] (10,20]
## [19] (10,20] (10,20]
## Levels: (1,10] (10,20]
table(y)
## y
## (1,10] (10,20]
## 9 10
# If changed the default behavior of include.lowest = TRUE

y<-cut(x, breaks = c(1,10,20), include.lowest = TRUE)

y
## [1] [1,10] [1,10] [1,10] [1,10] [1,10] [1,10] [1,10] [1,10] [1,10]
## [10] [1,10] (10,20] (10,20] (10,20] (10,20] (10,20] (10,20] (10,20] (10,20]
## [19] (10,20] (10,20]
## Levels: [1,10] (10,20]
table(y)
## y
## [1,10] (10,20]
## 10 10
```

*Example 2:* Recall from Chapter 11, that the geom_histogram() function plots a histogram with default 30 bins but the bins number can be changed.

```
library(tidyverse)
## -- Attaching packages ---------------------------------------------------------
-------- tidyverse 1.3.0 --
## <U+2713> ggplot2 3.2.1  <U+2713> purrr 0.3.3
## <U+2713> tibble 2.1.3  <U+2713> dplyr 0.8.3
## <U+2713> tidyr 1.0.0  <U+2713> stringr 1.4.0
## <U+2713> readr 1.3.1  <U+2713> forcats 0.4.0
## -- Conflicts -------------------------------------------------------------------
tidyverse_conflicts() --
## x dplyr::filter() masks stats::filter()
## x dplyr::lag() masks stats::lag()
library(compareGroups)
## Warning: package 'compareGroups' was built under R version 3.6.3
## Loading required package: SNPassoc
## Warning: package 'SNPassoc' was built under R version 3.6.3
## Loading required package: haplo.stats
## Warning: package 'haplo.stats' was built under R version 3.6.3
## Loading required package: survival
## Warning: package 'survival' was built under R version 3.6.3
## Loading required package: mvtnorm
## Loading required package: parallel
## Registered S3 method overwritten by 'SNPassoc':
## method   from
## summary.haplo.glm haplo.stats
data(regicor)

# the default 30 bins for age column

ggplot(data = regicor, aes(x = age))+geom_histogram()
## `stat_bin()` using `bins = 30`. Pick better value with `binwidth`.
```

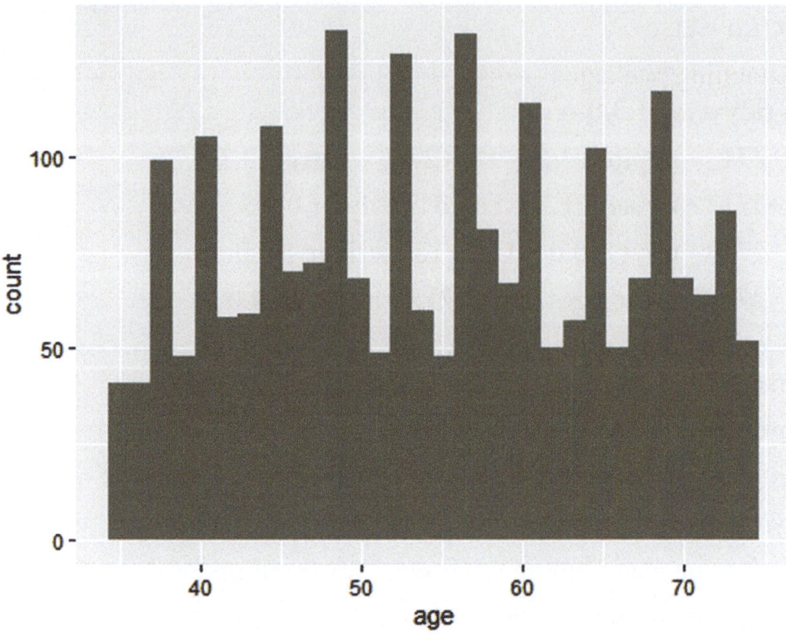


*# we can see these bins more clearly by setting red color to borders*

**ggplot**(data = regicor, **aes**(x = age))+**geom_histogram**(color = "red")
## `stat_bin()` using `bins = 30`. Pick better value with `binwidth`.

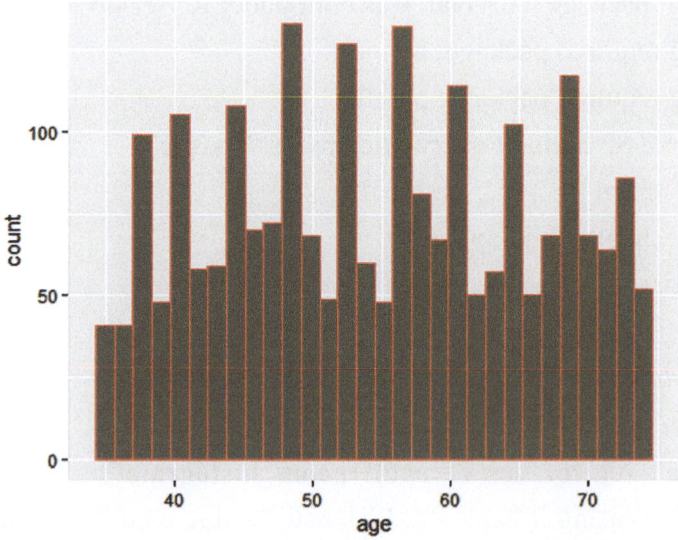


*# increase bins to 50*

**ggplot**(data = regicor, **aes**(x = age))+**geom_histogram**(color = "red", bins = 50)

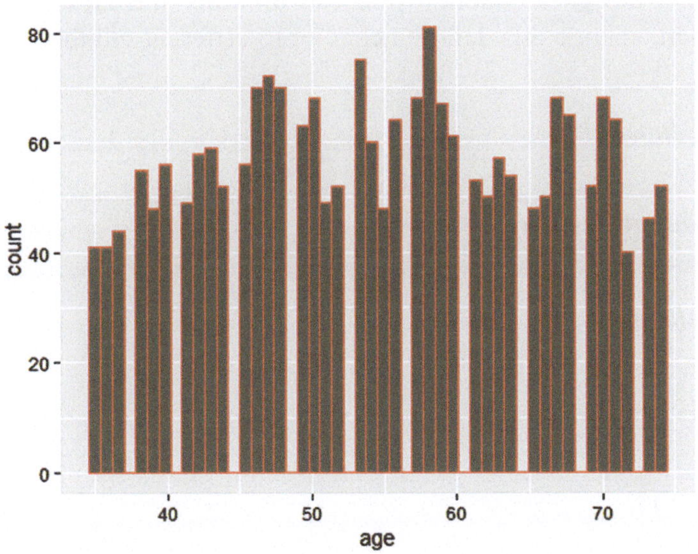


*# decrease bins to 20*

**ggplot**(data = regicor, **aes**(x = age))+**geom_histogram**(color = "red", bins = 20)

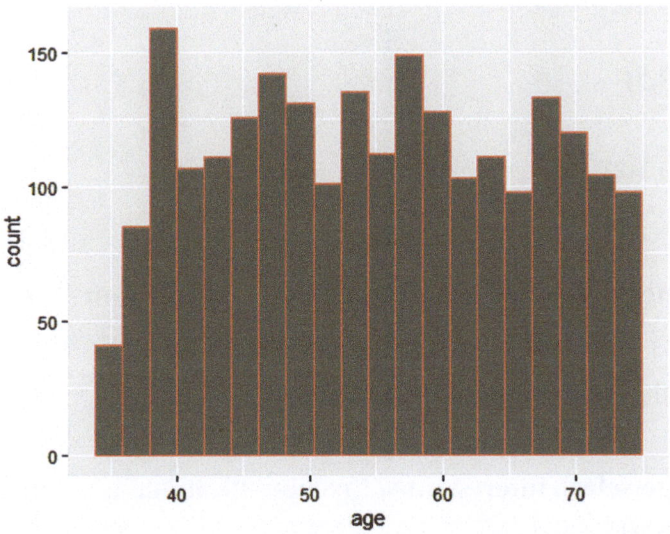

When you create a function in R, you can set a default argument and you can change it.

Example, design a function that can multiply the argument by 1 by default. Then, change this default behavior to different numbers.

```
#create the function

multiply<-function(x, by = 1){ x*by}

#default behavior

multiply(10)
## [1] 10
multiply(10,1)
## [1] 10
# change default

multiply(10,by=2)
## [1] 20
multiply(10,2)
## [1] 20
multiply(10,5)
## [1] 50
multiply(10,100)
## [1] 1000
```

### 12.1.3 Function as Arguments to Other Function

You can pass functions as arguments to another function.

Remember the example from Chapter 10.

```
dat<-arrange(select(filter(storms, name=="Katrina"), name,year,wind, pressure), desc(wind))
```

```
head(dat)
## # A tibble: 6 x 4
## name  year wind pressure
## <chr> <dbl> <int> <int>
## 1 Katrina 2005 150  902
## 2 Katrina 2005 145  909
## 3 Katrina 2005 140  905
## 4 Katrina 2005 125  930
## 5 Katrina 2005 125  913
## 6 Katrina 2005 110  920
```

Here we passed select() function as argument to arrange() function and also passed filter() function as argument to select() function.

The anonymous functions are non-named functions and they also can be passed to other functions.

*Example:* create a function (called evaluation) that returns the result of any applied function (func) on some data then pass different anonymous functions to the evaluation function.

```
# create evaluation function

evaluation<-function(func, dat) {x<-func(dat)

      x}

# use anonymous function that will return the first column of a matrix or dataframe

evaluation(function(x){x[,1]}, airquality)
## [1] 41 36 12 18 NA 28 23 19 8 NA 7 16 11 14 18 14 34 6
## [19] 30 11 1 11 4 32 NA NA NA 23 45 115 37 NA NA NA NA NA
```

```
## [37] NA 29 NA 71 39 NA NA 23 NA NA 21 37 20 12 13 NA NA NA
## [55] NA NA NA NA NA NA NA 135 49 32 NA 64 40 77 97 97 85 NA
## [73] 10 27 NA 7 48 35 61 79 63 16 NA NA 80 108 20 52 82 50
## [91] 64 59 39 9 16 78 35 66 122 89 110 NA NA 44 28 65 NA 22
## [109] 59 23 31 44 21 9 NA 45 168 73 NA 76 118 84 85 96 78 73
## [127] 91 47 32 20 23 21 24 44 21 28 9 13 46 18 13 24 16 13
## [145] 23 36 7 14 30 NA 14 18 20
evaluation(function(x){x[,1]}, state.x77)
## Alabama Alaska Arizona Arkansas California
## 3615 365 2212 2110 21198
## Colorado Connecticut Delaware Florida Georgia
## 2541 3100 579 8277 4931
## Hawaii Idaho Illinois Indiana Iowa
## 868 813 11197 5313 2861
## Kansas Kentucky Louisiana Maine Maryland
## 2280 3387 3806 1058 4122
## Massachusetts Michigan Minnesota Mississippi Missouri
## 5814 9111 3921 2341 4767
## Montana Nebraska Nevada New Hampshire New Jersey
## 746 1544 590 812 7333
## New Mexico New York North Carolina North Dakota Ohio
## 1144 18076 5441 637 10735
## Oklahoma Oregon Pennsylvania Rhode Island South Carolina
## 2715 2284 11860 931 2816
## South Dakota Tennessee Texas Utah Vermont
## 681 4173 12237 1203 472
## Virginia Washington West Virginia Wisconsin Wyoming
## 4981 3559 1799 4589 376
## use anonymous function to get the mean of the first column of a matrix or dataframe
```

**evaluation**(**function**(x){**mean**(x[,1], na.rm=TRUE)}, airquality)

## [1] 42.12931

**evaluation**(**function**(x){**mean**(x[,1], na.rm = TRUE)}, state.x77)

## [1] 4246.42

*## use anonymous function to get the number of missing values of a matrix or dataframe*

**evaluation**(**function**(x){**sum**(**is.na**(x))}, airquality)

## [1] 44

**evaluation**(**function**(x){**sum**(**is.na**(x))}, state.x77)

## [1] 0

**evaluation**(**function**(x){**sum**(**is.na**(x))}, regicor)

## [1] 1844

## 12.2 CONTROL STRUCTURES

Control structures in R allow you to control the flow of execution of different R expressions. Basically, control structures allow you to put some "logic" into your R code, rather than just always executing the same R code every time. Therefore, control structures allow you to respond to inputs or to features of the data and execute different R expressions accordingly.

Commonly used control structures are:

1. **If and Else:** Testing a condition and acting on it;
2. **For:** Execute a loop a fixed number of times;
3. **While:** Execute a loop while a condition is true;
4. **Repeat:** Execute an infinite loop (must break out of it to stop);
5. **Break:** Break the execution of a loop;
6. **Next:** Skip an iteration of a loop.

Most control structures are not used in interactive sessions, but when writing functions. The most used of these structures are the "if and else" and "for" control structures, so we will focus on them here.

### 12.1.1 If and else

The if-else combination is the most commonly used control structure in R. This structure allows you to test a condition and act on it depending on whether it's true or false.

you can just use the if statement

if() { do something }

The above code does nothing if the condition is false. If you have an action you want to execute when the condition is false, then you need an else clause.

if() { do something } else { do something else }

Alternatively, you can use the ifelse() function which have the following code:

ifelse(condition, do something, do something else)

Example, create a column for the dataframe airquality that is named "Temp_category" with levels "high" when Temp > 85 and "low or medium" otherwise.

```
library(datasets)

data("airquality")

head(airquality)
## Ozone Solar.R Wind Temp Month Day
## 1 41  190 7.4 67  5 1
## 2 36  118 8.0 72  5 2
## 3 12  149 12.6 74  5 3
## 4 18  313 11.5 62  5 4
## 5 NA  NA 14.3 56  5 5
## 6 28  NA 14.9 66  5 6
airquality$Temp_category<- ifelse(airquality$Temp > 85, "high", "low or medium")

head(airquality)
## Ozone Solar.R Wind Temp Month Day Temp_category
```

```
## 1 41  190 7.4 67  5 1 low or medium
## 2 36  118 8.0 72  5 2 low or medium
## 3 12  149 12.6 74  5 3 low or medium
## 4 18  313 11.5 62  5 4 low or medium
## 5 NA  NA 14.3 56  5 5 low or medium
## 6 28  NA 14.9 66  5 6 low or medium
```

**table**(airquality$Temp, airquality$Temp_category)

```
##
##  high low or medium
## 56 0    1
## 57 0    3
## 58 0    2
## 59 0    2
## 61 0    3
## 62 0    2
## 63 0    1
## 64 0    2
## 65 0    2
## 66 0    3
## 67 0    4
## 68 0    4
## 69 0    3
## 70 0    1
## 71 0    3
## 72 0    3
## 73 0    5
## 74 0    4
## 75 0    4
## 76 0    9
## 77 0    7
## 78 0    6
```

```
## 79 0   6
## 80 0   5
## 81 0  11
## 82 0   9
## 83 0   4
## 84 0   5
## 85 0   5
## 86 7   0
## 87 5   0
## 88 3   0
## 89 2   0
## 90 3   0
## 91 2   0
## 92 5   0
## 93 3   0
## 94 2   0
## 96 1   0
## 97 1   0
```

The table() function confirmed that the new column has values as we wish. You can also use mutate() function of dplyr package (a member of tidyverse) to create this column.

```
library(tidyverse)

data("airquality")

head(airquality)
## Ozone Solar.R Wind Temp Month Day
## 1 41  190 7.4 67  5 1
## 2 36  118 8.0 72  5 2
```

```
## 3 12  149 12.6 74  5 3
## 4 18  313 11.5 62  5 4
## 5 NA  NA 14.3 56  5 5
## 6 28  NA 14.9 66  5 6
airquality <- airquality %>% mutate(Temp_category = ifelse(airquality$Temp > 85, "high", "low or medium"))

head(airquality)
## Ozone Solar.R Wind Temp Month Day Temp_category
## 1 41  190 7.4 67  5 1 low or medium
## 2 36  118 8.0 72  5 2 low or medium
## 3 12  149 12.6 74  5 3 low or medium
## 4 18  313 11.5 62  5 4 low or medium
## 5 NA  NA 14.3 56  5 5 low or medium
## 6 28  NA 14.9 66  5 6 low or medium
table(airquality$Temp, airquality$Temp_category)
##
##  high low or medium
## 56 0   1
## 57 0   3
## 58 0   2
## 59 0   2
## 61 0   3
## 62 0   2
## 63 0   1
## 64 0   2
## 65 0   2
## 66 0   3
## 67 0   4
## 68 0   4
## 69 0   3
```

```
## 70 0  1
## 71 0  3
## 72 0  3
## 73 0  5
## 74 0  4
## 75 0  4
## 76 0  9
## 77 0  7
## 78 0  6
## 79 0  6
## 80 0  5
## 81 0 11
## 82 0  9
## 83 0  4
## 84 0  5
## 85 0  5
## 86 7  0
## 87 5  0
## 88 3  0
## 89 2  0
## 90 3  0
## 91 2  0
## 92 5  0
## 93 3  0
## 94 2  0
## 96 1  0
## 97 1  0
```

The table() function confirmed that the new column has values as we wish.

If you have a series of tests, you can use many nested ifelse() functions by following the code.

ifelse(condition, do something, ifelse(condition, do something, do something different)).

Example, create a column for the dataframe airquality that is named "Temp_category" with levels "high" when Temp > 85, "medium" when Temp > 79 and "low" otherwise.

```
library(datasets)

data("airquality")

head(airquality)
## Ozone Solar.R Wind Temp Month Day
## 1 41  190 7.4 67  5 1
## 2 36  118 8.0 72  5 2
## 3 12  149 12.6 74  5 3
## 4 18  313 11.5 62  5 4
## 5 NA  NA 14.3 56  5 5
## 6 28  NA 14.9 66  5 6
airquality$Temp_category<- ifelse(airquality$Temp > 85, "high",

        ifelse(airquality$Temp > 79, "medium", "low"))

head(airquality)
## Ozone Solar.R Wind Temp Month Day Temp_category
## 1 41  190 7.4 67  5 1   low
## 2 36  118 8.0 72  5 2   low
## 3 12  149 12.6 74  5 3   low
## 4 18  313 11.5 62  5 4   low
## 5 NA  NA 14.3 56  5 5   low
## 6 28  NA 14.9 66  5 6   low
table(airquality$Temp, airquality$Temp_category)
##
```

```
## high low medium
## 56 0 1 0
## 57 0 3 0
## 58 0 2 0
## 59 0 2 0
## 61 0 3 0
## 62 0 2 0
## 63 0 1 0
## 64 0 2 0
## 65 0 2 0
## 66 0 3 0
## 67 0 4 0
## 68 0 4 0
## 69 0 3 0
## 70 0 1 0
## 71 0 3 0
## 72 0 3 0
## 73 0 5 0
## 74 0 4 0
## 75 0 4 0
## 76 0 9 0
## 77 0 7 0
## 78 0 6 0
## 79 0 6 0
## 80 0 0 5
## 81 0 0 11
## 82 0 0 9
## 83 0 0 4
## 84 0 0 5
## 85 0 0 5
## 86 7 0 0
```

```
## 87 5 0  0
## 88 3 0  0
## 89 2 0  0
## 90 3 0  0
## 91 2 0  0
## 92 5 0  0
## 93 3 0  0
## 94 2 0  0
## 96 1 0  0
## 97 1 0  0
```

The table() function confirmed that the new column has values as we wish. The result of ifelse() control structure can also be created using the cut() function (see Chapter 6).

```
# summary statistics of Temp column to get its minimum and maximum

summary(airquality$Temp)
## Min. 1st Qu. Median Mean 3rd Qu. Max.
## 56.00 72.00 79.00 77.88 85.00 97.00
airquality$Temp_category2<-cut(airquality$Temp, breaks=c(55,79,85,97),
     labels = c("low","medium","high"))

head(airquality)
## Ozone Solar.R Wind Temp Month Day Temp_category Temp_category2
## 1 41  190 7.4 67  5 1   low   low
## 2 36  118 8.0 72  5 2   low   low
## 3 12  149 12.6 74  5 3   low   low
## 4 18  313 11.5 62  5 4   low   low
## 5 NA  NA 14.3 56  5 5   low   low
## 6 28  NA 14.9 66  5 6   low   low
```

```
table(airquality$Temp_category, airquality$Temp_category2)
##
##  low medium high
## high  0  0 34
## low  80  0 0
## medium 0  39 0
mean(airquality$Temp_category == airquality$Temp_category2)
## [1] 1
```

The table() and mean() functions confirmed that the new column (Temp_category2) has the same values as the old column (Temp_category).

### 12.1.2 For Loops

For loops are most commonly used for iterating over the elements of an object (list, vector, etc.).

Simple example:

```
for(i in 1:10) {
 print(i)
}
## [1] 1
## [1] 2
## [1] 3
## [1] 4
## [1] 5
## [1] 6
## [1] 7
## [1] 8
## [1] 9
## [1] 10
i
```

```
## [1] 10
```

This loop takes the i variable and in each iteration of the loop gives it values 1, 2, 3, …, 10, and then executes the code within the curly braces, and then the loop exits. If you print i, you will find that takes a value of 10 (last value).

Example 2: to calculate the mean of each column of airquality dataframe, we first create an empty vector called means, then use a for loop that iterates over numbers from 1 to 6 (airquality has 6 columns). In each iteration, the function will calculate the mean of that column number. Note that we use airquality[,i] to index for each column.

```
data("airquality")

means<-vector()

for (i in 1:6) {

 means[i]<-mean(airquality[,i], na.rm = TRUE)

 means}

means
## [1] 42.129310 185.931507 9.957516 77.882353 6.993464 15.803922
names(airquality)
## [1] "Ozone" "Solar.R" "Wind" "Temp" "Month" "Day"
names(means)<-names(airquality)

means
## Ozone Solar.R Wind Temp Month Day
## 42.129310 185.931507 9.957516 77.882353 6.993464 15.803922
## To confirm our results
```

```
mean(airquality[,1], na.rm = TRUE)
## [1] 42.12931
mean(airquality$Ozone, na.rm = TRUE)
## [1] 42.12931
mean(airquality$Day, na.rm = TRUE)
## [1] 15.80392
# there is also a function called colMeans()

colMeans(airquality, na.rm = TRUE)
## Ozone Solar.R Wind Temp Month Day
## 42.129310 185.931507 9.957516 77.882353 6.993464 15.803922
```

## 12.3 LOOP FUNCTIONS

Each of the apply functions or loop functions split some data into smaller pieces, apply a function to each piece then combine the results. This is called split-apply-combine strategy for data analysis.

### 12.3.1 Lapply

lapply() function takes a list (or a dataframe as it is a list of different columns) as input, applies a function to each element, and returns the result as a list that has the same length as the input list.

Example, use lapply() to get a useful function result of every column and other arguments of the used function (as na.rm = TRUE) can be applied as arguments to lapply.

```
data("airquality")

# to get standard deviation of every column

sd_result<-lapply(airquality, sd)

sd_result
## $Ozone
```

```
## [1] NA
##
## $Solar.R
## [1] NA
##
## $Wind
## [1] 3.523001
##
## $Temp
## [1] 9.46527
##
## $Month
## [1] 1.416522
##
## $Day
## [1] 8.86452
class(sd_result)
## [1] "list"
length(airquality)
## [1] 6
length(sd_result)
## [1] 6
sd_result<-lapply(airquality, sd, na.rm=TRUE)

sd_result
## $Ozone
## [1] 32.98788
##
## $Solar.R
## [1] 90.05842
##
```

```
## $Wind
## [1] 3.523001
##
## $Temp
## [1] 9.46527
##
## $Month
## [1] 1.416522
##
## $Day
## [1] 8.86452
# to get class of every column

class_result<-lapply(airquality, class)

class_result
## $Ozone
## [1] "integer"
##
## $Solar.R
## [1] "integer"
##
## $Wind
## [1] "numeric"
##
## $Temp
## [1] "integer"
##
## $Month
## [1] "integer"
##
```

```
## $Day
## [1] "integer"
# to get sum of every column

sum_result<-lapply(airquality, sum, na.rm=TRUE)

sum_result
## $Ozone
## [1] 4887
##
## $Solar.R
## [1] 27146
##
## $Wind
## [1] 1523.5
##
## $Temp
## [1] 11916
##
## $Month
## [1] 1070
##
## $Day
## [1] 2418
# to get summary statistics of every column

summary_result<-lapply(airquality, summary)

summary_result
## $Ozone
## Min. 1st Qu. Median Mean 3rd Qu. Max. NA's
```

```
## 1.00 18.00 31.50 42.13 63.25 168.00  37
##
## $Solar.R
## Min. 1st Qu. Median Mean 3rd Qu. Max. NA's
##  7.0 115.8 205.0 185.9 258.8 334.0  7
##
## $Wind
## Min. 1st Qu. Median Mean 3rd Qu. Max.
## 1.700 7.400 9.700 9.958 11.500 20.700
##
## $Temp
## Min. 1st Qu. Median Mean 3rd Qu. Max.
## 56.00 72.00 79.00 77.88 85.00 97.00
##
## $Month
## Min. 1st Qu. Median Mean 3rd Qu. Max.
## 5.000 6.000 7.000 6.993 8.000 9.000
##
## $Day
## Min. 1st Qu. Median Mean 3rd Qu. Max.
##  1.0  8.0 16.0 15.8 23.0 31.0
```

Notice that the length of airquality (input list) is the same as the length of sd_result (output list).

### 12.3.2 Sapply

sapply() function is the same as lapply() in taking a list as input and applying a function to each element, but it returns the result as a vector that has the same length as the input list.

This occurs if each element of the result contains only one value (as sum, mean, sd). However, if each element of the result contains multiple values (as summary), sapply() will return a list as lapply.

```
data("airquality")

# to get standard deviation of every column

sd_result<-sapply(airquality, sd, na.rm=TRUE)

sd_result
## Ozone Solar.R Wind Temp Month Day
## 32.987885 90.058422 3.523001 9.465270 1.416522 8.864520
class(sd_result)
## [1] "numeric"
length(airquality)
## [1] 6
length(sd_result)
## [1] 6
# to get summary statistics of every column

summary_result<-sapply(airquality, summary)

summary_result
## $Ozone
## Min. 1st Qu. Median Mean 3rd Qu. Max. NA's
## 1.00 18.00 31.50 42.13 63.25 168.00  37
##
## $Solar.R
## Min. 1st Qu. Median Mean 3rd Qu. Max. NA's
##  7.0 115.8 205.0 185.9 258.8 334.0  7
##
## $Wind
## Min. 1st Qu. Median Mean 3rd Qu. Max.
## 1.700 7.400 9.700 9.958 11.500 20.700
```

```
##
## $Temp
## Min. 1st Qu. Median Mean 3rd Qu. Max.
## 56.00 72.00 79.00 77.88 85.00 97.00
##
## $Month
## Min. 1st Qu. Median Mean 3rd Qu. Max.
## 5.000 6.000 7.000 6.993 8.000 9.000
##
## $Day
## Min. 1st Qu. Median Mean 3rd Qu. Max.
##  1.0  8.0 16.0 15.8 23.0 31.0
class(summary_result)
## [1] "list"
```

### 12.3.3 Tapply

tapply() takes a vector and splits it by a factor then applies a function to each subset of that vector and returns the result. Note that length of the input vector and the input factor must be the same.

Example, define the mean temperature for every month of the airquality dataframe:

```
data("airquality")

length(airquality$Temp)
## [1] 153
length(airquality$Month)
## [1] 153
# get the mean of each subset
```

```
mean_result<-tapply(airquality$Temp, airquality$Month, mean, na.rm =
TRUE)

mean_result
## 5 6 7 8 9
## 65.54839 79.10000 83.90323 83.96774 76.90000
# get the summary statistics of each subset

summary_result<-tapply(airquality$Temp, airquality$Month, summary)

summary_result
## $`5`
## Min. 1st Qu. Median Mean 3rd Qu. Max.
## 56.00 60.00 66.00 65.55 69.00 81.00
##
## $`6`
## Min. 1st Qu. Median Mean 3rd Qu. Max.
## 65.00 76.00 78.00 79.10 82.75 93.00
##
## $`7`
## Min. 1st Qu. Median Mean 3rd Qu. Max.
## 73.0 81.5 84.0 83.9 86.0 92.0
##
## $`8`
## Min. 1st Qu. Median Mean 3rd Qu. Max.
## 72.00 79.00 82.00 83.97 88.50 97.00
##
## $`9`
## Min. 1st Qu. Median Mean 3rd Qu. Max.
## 63.0 71.0 76.0 76.9 81.0 93.0
```

The same results can be obtained by split() and sapply() functions. The split() function is discussed in Chapter 6.

```
l<-split(airquality$Temp, airquality$Month)

l
## $`5`
## [1] 67 72 74 62 56 66 65 59 61 69 74 69 66 68 58 64 66 57 68 62 59 73 61 61 57
## [26] 58 57 67 81 79 76
##
## $`6`
## [1] 78 74 67 84 85 79 82 87 90 87 93 92 82 80 79 77 72 65 73 76 77 76 76 76 75
## [26] 78 73 80 77 83
##
## $`7`
## [1] 84 85 81 84 83 83 88 92 92 89 82 73 81 91 80 81 82 84 87 85 74 81 82 86 85
## [26] 82 86 88 86 83 81
##
## $`8`
## [1] 81 81 82 86 85 87 89 90 90 92 86 86 82 80 79 77 79 76 78 78 77 72 75 79 81
## [26] 86 88 97 94 96 94
##
## $`9`
## [1] 91 92 93 93 87 84 80 78 75 73 81 76 77 71 71 78 67 76 68 82 64 71 81 69 63
## [26] 70 77 75 76 68
sapply(l,mean, na.rm=TRUE)
## 5 6 7 8 9
```

```
## 65.54839 79.10000 83.90323 83.96774 76.90000
sapply(l, summary)
##   5 6 7 8 9
## Min. 56.00000 65.00 73.00000 72.00000 63.0
## 1st Qu. 60.00000 76.00 81.50000 79.00000 71.0
## Median 66.00000 78.00 84.00000 82.00000 76.0
## Mean 65.54839 79.10 83.90323 83.96774 76.9
## 3rd Qu. 69.00000 82.75 86.00000 88.50000 81.0
## Max. 81.00000 93.00 92.00000 97.00000 93.0
```

Also this result can also be obtained using group_by() and summarise() functions of the tidyverse package.

```
library(tidyverse)

airquality %>% group_by(Month) %>% summarise(mean_temp = mean(Temp, na.rm = TRUE))
## # A tibble: 5 x 2
## Month mean_temp
## <int> <dbl>
## 1 5 65.5
## 2 6 79.1
## 3 7 83.9
## 4 8 84.0
## 5 9 76.9
```

### 12.3.4 Apply()

The apply() function is used to a evaluate a function over the margins (1 for rows and 2 for columns) of a dataframe or matrix.

Example, calculate the row means and column means of airquality dataframe.

```
data("airquality")
```

**dim**(airquality)

## [1] 153 6

*# get the row means, there are 153 rows so we get 153 means*

res<-**apply**(airquality,1,mean, na.rm = TRUE)

res

## [1] 51.90000 40.16667 42.60000 68.91667 20.07500 23.98000 67.93333 33.96667

## [9] 20.35000 57.32000 20.78000 61.28333 65.70000 64.31667 29.03333 74.08333

## [17] 73.50000 30.40000 75.91667 25.28333 17.28333 74.60000 21.28333 37.66667

## [25] 33.92000 73.98000 24.25000 24.66667 71.15000 76.28333 72.56667 75.92000

## [33] 75.74000 66.82000 57.84000 64.92000 73.86000 43.45000 76.18000 80.13333

## [41] 79.41667 75.98000 73.84000 46.66667 89.16000 86.70000 54.31667 72.78333

## [49] 25.86667 40.25000 43.71667 52.06000 32.94000 40.12000 72.46000 49.80000

## [57] 49.00000 32.66000 44.70000 31.58000 53.00000 83.35000 66.70000 61.36667

## [65] 41.38000 56.43333 76.81667 76.68333 79.55000 80.45000 62.23333 49.52000

## [73] 63.38333 52.98333 83.58000 28.55000 69.81667 70.88333 76.88333 64.01667

## [81] 67.75000 21.98333 75.54000 83.70000 83.26667 76.00000 37.43333 44.33333

## [89] 70.90000 75.73333 74.06667 73.53333 36.48333 22.96667 32.23333 36.58000

## [97] 28.08000 34.32000 80.83333 72.38333 72.00000 68.12000 50.70000 58.91667

## [105] 69.25000 55.61667 35.50000 34.05000 36.71667 41.23333 65.15000 58.38333

## [113] 66.91667 26.88333 74.72000 62.95000 87.23333 69.33333 56.34000 70.28333

## [121] 79.38333 76.88333 68.71667 61.81667 63.85000 60.63333 65.10000 41.73333

## [129] 39.75000 63.15000 58.05000 59.15000 64.11667 65.98333 65.41667 61.88333

## [137] 22.98333 38.58333 65.48333 58.13333 25.55000 61.38333 56.00000 59.60000

## [145] 24.70000 49.71667 28.05000 24.60000 55.81667 54.24000 55.21667 45.16667

## [153] 60.25000

*# get the column means, there are 6 columns so we get 6 means*

res<-**apply**(airquality,2,mean, na.rm = TRUE)

res

## Ozone Solar.R Wind Temp Month Day

## 42.129310 185.931507 9.957516 77.882353 6.993464 15.803922

Example, get the row means and column means of state.x77 matrix.

**dim**(state.x77)

## [1] 50 8

*# get the row means, there are 50 rows so we get 50 means*

res<-**apply**(state.x77, 1, mean, na.rm = TRUE)

res

## Alabama Alaska Arizona Arkansas California

## 7261.819 71676.601 15039.031 7202.570 22854.839

```
## Colorado Connecticut Delaware Florida Georgia
## 13937.558 1697.710 950.595 8416.032 8410.005
## Hawaii Idaho Illinois Indiana Iowa
## 1549.950 10984.034 9039.118 5765.198 7963.045
## Kansas Kentucky Louisiana Maine Maryland
## 11123.448 5870.600 6552.495 4495.186 2443.115
## Massachusetts Michigan Minnesota Mississippi Missouri
## 2329.091 8867.429 11022.308 6613.624 9781.699
## Montana Nebraska Nevada New Hampshire New Jersey
## 18871.295 10351.175 14495.279 1803.354 2541.966
## New Mexico New York North Carolina North Dakota Ohio
## 15801.802 8878.444 7289.326 9413.535 7065.903
## Oklahoma Oregon Pennsylvania Rhode Island South Carolina
## 9461.565 12913.616 7691.091 848.375 4607.583
## South Dakota Tennessee Texas Utah Vermont
## 10137.823 6189.576 34840.838 10950.413 1743.605
## Virginia Washington West Virginia Wisconsin Wyoming
## 6209.472 9395.640 3713.148 7975.085 12807.336
# get the column means, there are 8 columns so we get 8 means

res<-apply(state.x77, 2, mean, na.rm = TRUE)

res
## Population Income Illiteracy Life Exp Murder HS Grad Frost
## 4246.4200 4435.8000 1.1700 70.8786 7.3780 53.1080 104.4600
## Area
## 70735.8800
```

# BIBLIOGRAPHY


1. Anderson, E., (1935). The irises of the Gaspe Peninsula. *Bulletin of the American Iris Society, 59*, 2–5.
2. Ayres et al., (2000). SisPorto 2.0: A program for automated analysis of cardiotocograms. *J. Matern. Fetal. Med., 5*, 311–318.
3. Chambers, J. M., Cleveland, W. S., Kleiner, B., & Tukey, P. A., (1983). *Graphical Methods for Data Analysis*. Belmont, CA: Wadsworth.
4. Garrett, G., & Hadley, W., (2011). Dates and times made easy with lubridate. *Journal of Statistical Software, 40*(3), 1–25. http://www.jstatsoft.org/v40/i03/ (accessed on 25 March 2021).
5. Hadley, W., & Dana, S., (2019). *Scales: Scale Functions for Visualization*. R package version 1.1.0. https://CRAN.R-project.org/package=scales (accessed on 25 March 2021).
6. Hadley, W., & Jennifer, B., (2019). *readxl: Read Excel Files*. R package version 1.3.1. https://CRAN.R-project.org/package=readxl (accessed on 25 March 2021).
7. Hadley, W., & Lionel, H., (2019). *tidyr: Tidy Messy Data*. R package version 1.0.0. https://CRAN.R-project.org/package=tidyr (accessed on 25 March 2021).
8. Hadley, W., (2019). *nycflights13: Flights that Departed NYC in 2013*. R package version 1.0.1. https://CRAN.R-project.org/package=nycflights13 (accessed on 25 March 2021).
9. Hadley, W., Romain, F., Lionel, H., & Kirill, M., (2019). *dplyr: A Grammar of Data Manipulation*. R package version 0.8.3. https://CRAN.R-project.org/package=dplyr (accessed on 25 March 2021).

10. Isaac, S., Hector, S., & Joan, V., (2014). Building bivariate tables: The compare groups package for R. *Journal of Statistical Software, 57*(12), 1–16. http://www.jstatsoft.org/v57/i12/ (accessed on 25 March 2021).
11. Moro, S., Cortez, P., & Rita, P., (2014). A data-driven approach to predict the success of bank telemarketing. *Decision Support Systems, 62*, 22–31, Elsevier. Available at: http://media.salford-systems.com/video/tutorial/2015/targeted_marketing.pdf.
12. R Core Team, (2019). *R: A Language and Environment for Statistical Computing*. R Foundation for Statistical Computing, Vienna, Austria. https://www.R-project.org/ (accessed on 25 March 2021).
13. Rafael, A. I., & Amy, G., (2019). *dslabs: Data Science Labs*. R package version 0.7.3. https://CRAN.Rproject.org/package=dslabs (accessed on 25 March 2021).
14. Roger, D. P., (2016). *Exploratory Data Analysis with R*. http://leanpub.com/exdata (accessed on 25 March 2021).
15. Roger, D. P., (2016). *R Programming for Data Science*. http://leanpub.com/rprogramming (accessed on 25 March 2021).
16. Roger, D. P., Sean, K., & Brooke, A., (2017). *Mastering Software Development in R*. http://leanpub.com/msdr (accessed on 25 March 2021).
17. Stefan, M. B., & Hadley, W., (2014). *magrittr: A Forward-Pipe Operator for R*. R package version 1.5. https://CRAN.R-project.org/package=magrittr (accessed on 25 March 2021).
18. Wickham, et al., (2019). Welcome to the tidy verse. *Journal of Open Source Software, 4*(43), 1686, https://doi.org/10.21105/joss.01686.
19. Wickham, H., (2016). *ggplot2: Elegant Graphics for Data Analysis*. Springer-Verlag New York.
20. Yihui, X., (2019). *knitr: A General-Purpose Package for Dynamic Report Generation in R*. R package version 1.26.

# Index

## V



## W



## Y